Current Chinese Economic Report Series

The Current Chinese Economic Reports series provides insights into the economic development of one of the largest and fastest growing economies in the world; though widely discussed internationally, many facets of its current development remain unknown to the English speaking world. All reports contain new data, which was previously unknown or unavailable outside of China. The series covers regional development, industry reports, as well as special topics like environmental or demographical issues.

More information about this series at http://www.springer.com/series/11028

Zhongguancun Listed Companies Association

The Growth Report of Zhongguancun NEEQ Listed Companies (2020)

Springer

Zhongguancun Listed Companies Association
Zhongguancun Listed Companies Association
Beijing, China

ISSN 2194-7937 ISSN 2194-7945 (electronic)
Current Chinese Economic Report Series
ISBN 978-981-33-6818-7 ISBN 978-981-33-6819-4 (eBook)
https://doi.org/10.1007/978-981-33-6819-4

This Springer imprint is published by the registered company Springer Nature Singapore Pte Ltd.
The registered company address is: 152 Beach Road, #21-01/04 Gateway East, Singapore 189721, Singapore

Report Instruction

The Growth Report of Zhongguancun NEEQ Listed Companies: The Report was drafted and edited by the NEEQ Branch of Zhongguancun Listed Companies Association (hereinafter referred to as ZLCA), focusing on the research and study of Zhongguancun NEEQ listed companies. *The Growth Report of Zhongguancun NEEQ Listed Companies (2020)*, issued in 2020, refers to the report compiled based on the annual reports of Zhongguancun NEEQ companies in 2019.

Objects of Research: Objects of research include enterprises registered in Beijing, and listed on the National Equities Exchange and Quotations (NEEQ), excluding companies listed on the STAQ system and NET system, and main board delisted companies.

Data Sources: All financial data without special remarks was collected from the WIND database, and the patent-related data was quoted from Zhongguancun Intellectual Property Promotion Office. Besides, part of the comparative data was cited from the official website of NEEQ and Beijing Municipal Bureau of Statistics, as well as "The Competitiveness Report of Zhongguancun Listed Companies".

Methodology: A database was created by assembling together all the public financial data of Zhongguancun NEEQ listed companies. Based on the database, the overall growth, distribution and status of diverse industries were analysed vertically and horizontally. Conclusions were then drawn from the analysis and development suggestions were proposed.

Currency Unit: RMB was used as the currency unit throughout this report.

Data Check: The financial data quoted in this report was collected and checked by the project team with valuable guidance and assistance from Ernst & Young Hua Ming LLP.

About the Author

Zhongguancun Listed Companies Association Zhongguancun Listed Companies Association (ZLCA), founded autonomously by a number of listed companies in Zhongguancun, is a national NGO registered under the Bureau of Civil Affairs of Beijing. The association was officially established on August 16th, 2012.

ZLCA's mission is to assist the government in delivering and implementing policies, representing members in providing advices and suggestions to the government, providing professional policy consultation for its members and establishing a high-level networking, resource sharing and learning platform.

The foundation of ZLCA will play an important role in helping listed companies operate normatively and develop healthily, building a bridge between listed companies, investors, capital market supervisory bodies and government, promoting sharing of resources and cooperation.

Objectives

Observation of Authoritative Data

Through the analysis of growth data, the study of growth pattern and the publication of the annual competitiveness report of ZLCs, this association can grasp the development of ZLCs and aims to assist the government, members and investors in their relevant project research and development strategies for member companies.

Communication Bridge

We are committed to the communication between companies, investors, market regulators and the government, improve their mutual understanding by respecting their needs and views.

Promotion for Innovation

We strive to facilitate the development of initiative companies. We also wish to support companies listed on Growth Companies Market and promote technological innovation in Zhongguancun.

International Influence

We will host international conferences, activities and organise overseas roadshows with group visits to boost the international influence of companies in Zhongguancun.

Introduction

The NEEQ is positioned to serve innovative, entrepreneurial and growth-oriented small, medium and micro enterprises, and plays an important role in serving the private economy and the development of small and medium enterprises. Zhongguancun, as the birthplace of the NEEQ market, the sustained, healthy and stable development of the enterprises in the park will have a certain demonstrative effect on the NEEQ enterprises across the country. Therefore, research on Zhongguancun NEEQ enterprises will be conducive to explore the capital market to serve technological innovation-oriented small, medium and micro enterprises, and is of great significance for further exerting the boosting role of the capital market in technological innovation.

As of December 31, 2019, the number of companies listed on the NEEQ in Zhongguancun was 1,190, accounting for 13.29% of the total number of NEEQ companies in the country. Ranked second in the number of companies on the National New Third Board, second only to Guangdong Province. Among them, there are 90 innovative enterprises in Zhongguancun NEEQ ranking first in the country. From the perspective of market value, the total market value of Zhongguancun NEEQ enterprises in 2019 was 340.05 billion yuan, and the total market value ranked first in the country, ahead of other regions. From the perspective of operating conditions, due to the decrease in the number of listed companies, the total operating income and net profit of Zhongguancun NEEQ companies have shown a downward trend, but the continuous operation of Zhongguancun NEEQ companies have outstanding performance, with 1035 continuous operations from 2018 to 2019. The company's operating income and net profit in 2019 were 205.121 billion yuan and 8.753 billion yuan, an increase of 6.53% and 21.95% year-on-year, respectively, indicating that the overall scale and quality of continuing operations are continuously improving. From the perspective of R&D, among the Zhongguancun NEEQ companies that disclosed R&D expenses in 2019, 368 companies were high-intensity R&D companies, accounting for 42.7%, an increase of 2.7% year-on-year, indicating that the decline in macro fundamentals did not affect the company's performance. Research and development process. The leading enterprises of Zhongguancun NEEQ have increased their R&D investment by contrarian trends and improved their core competitiveness. From the perspective of financing, Zhongguancun NEEQ listed companies actually raised 3.740 billion yuan through private placement and 5.510 billion yuan through bond issuance. Under the situation of inactive bond financing in the overall NEEQ market, the bond financing of Zhongguancun NEEQ companies performance is also better than other regions. However, the scale of corporate financing is still not optimistic, and financing support needs to be increased. From the perspective of regional comparison, in 2019, under the background of the overall macroeconomic downturn, the overall performance of the NEEQ companies in each major region has shown a downward trend, but the overall performance decline is mainly due to the decrease in the number of companies rather than the performance of continuing operations. However, Z—Park still has a strong competitive advantage compared to other regions.

Overall, in 2019, affected by the macroeconomic situation, the overall operating performance of Zhongguancun and even the NEEQ companies across the country will be affected to a certain extent, but there are many high-growth companies with stable operating conditions, continuous growth in performance and increased R&D investment. However, we still have to pay attention to the difficulties of financing, expensive financing and low innovation quality faced by Zhongguancun NEEQ enterprises in the development process.

In response to the financing difficulties and expensive financing problems faced by Zhongguancun NEEQ in the development process, we started with the expansion of direct financing and indirect financing channels. In terms of broadening the access to direct financing, (1) It is recommended that the national SME share transfer system fully seize the opportunity of the NEEQ reform, establish and improve the market mechanism for survival of the fittest, improve the overall quality of the NEEQ enterprises and then enhance the NEEQ market liquidity and improve enterprises. The proportion of direct financing and the efficiency of financing will effectively serve the innovative, entrepreneurial and growing small, medium and micro enterprises. (2) It is recommended that the Beijing Municipal Government set up a special policy fund to invest in enterprises with high technology content and great market potential, but with temporary difficulties in operation, and to support the development of Zhongguancun NEEQ enterprises through policies. From the aspect of expanding indirect financing channels, (1) It is recommended to subsidize the financial products of non-governmental organizations in an appropriate proportion. (2) It is recommended that the government encourages the enterprises to innovate financing models. (3) It is recommended that relevant government departments actively promote the practical implementation of inclusive financial support measures for small, medium and micro enterprises to solve the problem of difficult financing for small, medium and micro enterprises.

In response to the problem of low innovation quality, (1) It is suggested that Zhongguancun NEEQ listed companies can rely on external resources such as universities, scientific research institutions and technical service institutions to further enhance their innovation capabilities and improve the quality of their innovation output; (2) The relevant Beijing government can reduce unnecessary applications and approval procedures and speed up the progress of patent approval, so that enterprises can enjoy the benefits of R&D as soon as possible and stimulate their innovation vitality.

In view of the high actual tax burden of Zhongguancun NEEQ companies, it is recommended that relevant government departments give a certain degree to strategic emerging industries such as energy conservation and environmental protection, new generation information technology, biology, high-end equipment manufacturing, new energy, new materials and new energy vehicles, guiding the development and growth of such industries and forming an industry cluster effect in the Zhongguancun area.

Keywords: Zhongguancun, New Third Board, Growth capacity

Editorial Board of *The Growth Report of Zhongguancun NEEQ Listed Firms*

Contents

Chapter 1
Development Characteristics and Suggestions of Zhongguancun NEEQ Companies

Abstract The NEEQ market is an important link in the construction of a multi-level capital market, and it plays an important role in establishing financing channels for SMEs and improving the macroeconomic system. This report takes Zhongguancun NEEQ enterprises as the main research body. It systematically analyzes the number of NEEQ enterprises, capital market performance, profitability, debt issuance, innovation capacity, and the actual tax burden borne by enterprises, and objectively presents Zhongguancun NEEQ enterprises's growth characteristics and existing problems. The data shows that due to the impact of the economic downturn and the increase in the number of delisted companies, the overall development status of Zhongguancun NEEQ enterprises in 2019 is not optimistic, but the performance of continuing operating enterprises is relatively stable, and the corporate tax burden is significantly reduced. Deepening reform of the NEEQ brings new opportunities for enterprises.

Keyword Zhongguancun NEEQ · Market liquidity · R&D input

1.1 Development Characteristics of Zhongguancun NEEQ Market in 2019

1.1.1 Market Declining Slows Down, Ushering in New Opportunities for Reform

1.1.1.1 Number of Zhongguancun Listed Companies

As of December 31st, 2019, the number of Zhongguancun NEEQ listed companies (ZLCs) was 1,190. This report excludes companies that were delisted[1] and those

[1] The delisted enterprises here refer to the NEEQ enterprises of Zhongguancun which were delisted before December 31st, 2019.

Zhongguancun Listed Companies Association, *The Growth Report of Zhongguancun NEEQ Listed Companies (2020)*, Current Chinese Economic Report Series,
https://doi.org/10.1007/978-981-33-6819-4_1

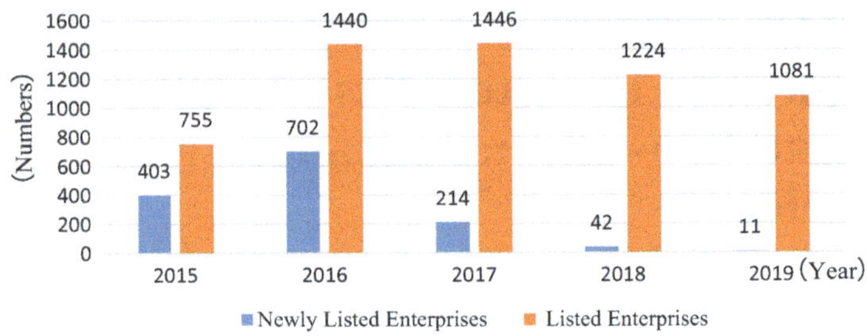

Fig. 1.1 2015–2019 Zhongguancun NEEQ listed companies and newly listed companies. *Source* Wind, compiled by ZLCA

Table 1.1 2018–2019 classification of Zhongguancun NEEQ listed companies

Year	Total	Innovation		Base	
	Numbers	Numbers	Percentage (%)	Numbers	Percentage (%)
2018	1224	122	9.97	1102	90.03
2019	1081	90	8.33	991	91.67

Source Wind, compiled by ZLCA

not disclosing their annual reports[2] as well as companies with other special conditions. The report only takes as research object the 1,081 Zhongguancun NEEQ listed companies that publicly disclosed their 2019 annual reports before June 30th, 2019.[3]

The number of Zhongguancun NEEQ listed companies has shown a downward trend for two consecutive years, but the decline rate has slowed down. From the perspective of the total number of listed companies, at the end of 2019, there were 1,081 Zhongguancun NEEQ listed companies, a year-on-year decrease of 11.68%, which was lower than the 15.35% decrease rate in 2018. The number of newly added NEEQ listed companies in Z-Park (Zhongguancun Science Park) in 2019 was 11, decreasing by 31 compared with the same period of last year (see Fig. 1.1).

Zhongguancun is the top performer in the country in terms of the number of NEEQ listed companies. From the angle of market classification,[4] the number of enterprises on the NEEQ Innovation has declined in 2019, but remained the first in the country. As of the end of 2019, the number of Zhongguancun NEEQ listed companies on Innovation tier was 90, accounting for 8.33% of the Zhongguancun NEEQ listed enterprises, a slight decrease from 9.97% in 2018 (see Table 1.1). From a nationwide

[2]The enterprises not disclosing annual report here refer to NEEQ enterprises of Zhongguancun that have not disclosed the annual report in 2019 before June 30th, 2020.

[3]In April 8th, 2020, NEEQ issued an announcement on the audit and disclosure of the 2019 annual report of listed companies. The announcement clearly stipulates that listed companies unable to disclose the audited annual report by April 30th because of the Covid-19 pandemic are allowed to delay the disclosure but it will not be later than June 30th in principle.

[4]The data of companies on Innovation tier is based on the December 31st source from Wind of the number of NEEQ enterprises belong to Innovation tier.

angle, in 2019, the number of Zhongguancun NEEQ listed companies on Innovative tier ranked the first in the country, accounting for 13.10% of the total on this tier, outnumbering other regions of Guangdong, Jiangsu, Zhejiang, and Shanghai.

1.1.1.2 Number of Delisted Enterprises

The increase in the number of Zhongguancun NEEQ delisting enterprises was much lower than that of the whole country. In 2019, a total of 1,987 NEEQ listed companies across the country were delisted, and the number of delisting increased by 30.98% compared with 2018. The increase was much lower than the 113.96% in 2018. Among them, Zhongguancun had a total of 258 delisting enterprises in 2019 with a year-on-year increase of 14.67%. This increase was also much lower than the 161.63% in 2018 (see Fig. 1.2).

The report has made a detailed analysis of the reasons behind the delisting in 2019. 57 enterprises were delisted for not disclosing regular reports after suspension of listing, accounting for 22.09%; 9 enterprises were delisted because of backdoor listing, accounting for 3.49%; and 102 other enterprises were delisted because they did not meet the listing requirements, accounting for 39.53%; 90 enterprises were delisted due to adjustments in production and operation, accounting for 34.88% (see Fig. 1.3). Among the 9 enterprises that were delisted due to backdoor listing, 7 have been successfully listed on the A-share market; 2 have terminated their IPOs. As of the end of 2019, a total of 21 Zhongguancun NEEQ listed companies have successfully been listed on the A-share market, indicating that there are some high-quality Zhongguancun NEEQ listed companies (see Table 1.2).

On the whole, in recent years, both the national NEEQ market and the Zhongguancun NEEQ market have shown that the number of listing is decreasing year by year, while the number of delisting has increased at the same time. It is mainly because of the following problems existed before the comprehensive deepening of the NEEQ reform.

First, it is difficult to refinance on the NEEQ market. At present, small, micro and medium-sized enterprises in China are generally facing the difficulties of "accessing

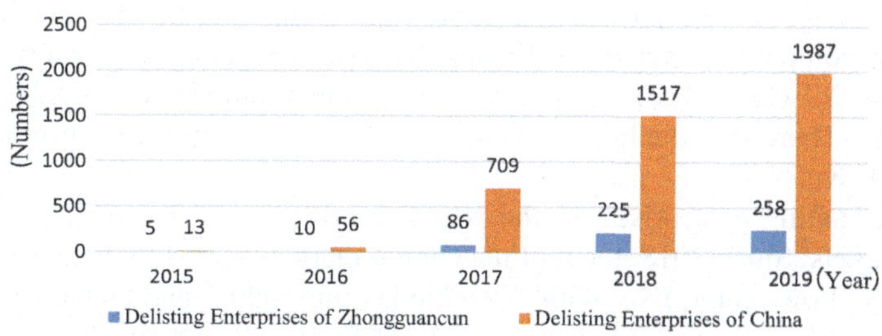

Fig. 1.2 Statistics of delisting companies of China and Zhongguancun from 2015 to 2019. *Source* Wind, compiled by ZLCA

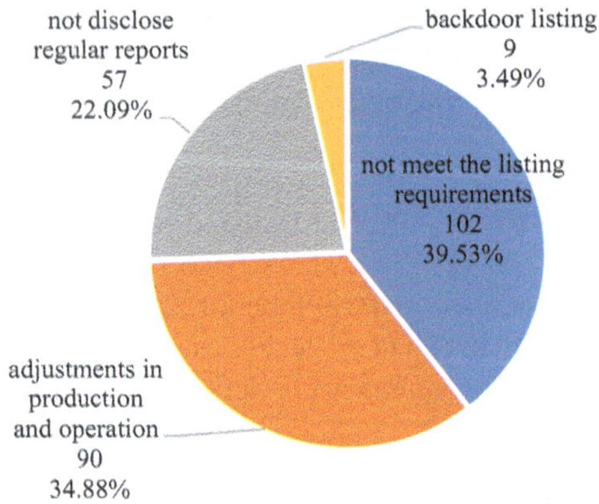

Fig. 1.3 Statistics of delisting reasons of Zhongguancun NEEQ delisted enterprises in 2019. *Source* Wind, compiled by ZLCA

to affordable financing". The launch of NEEQ market has undoubtedly provided new financing channels for SMEs (small and medium-sized enterprises), effectively eased the financing woes of some enterprises, and helped some develop and grow via NEEQ market. However, most NEEQ listed companies can only enjoy the financing dividends at the time of their first listing. The difficulty for enterprises to refinance through the NEEQ market has been increasing year by year. From a national perspective, between 2015 and 2019 NEEQ listed firms that launched equity private placement accounted for 31.41%, 16.29%, 12.07%, 7.61%, and 5.38% respectively. The number of Zhongguancun NEEQ listed companies that implemented private placements from 2015 to 2019 accounted for 33.94%, 18.42%, and 12.05%, 7.29%, 4.87% respectively.

Second, the NEEQ market lacks liquidity. In 2016, the expansion of the NEEQ market led to a mix of good and bad enterprises get listed on the NEEQ, and the high threshold for qualified investors getting into the NEEQ market caused a significant decline in the liquidity of the NEEQ market. From 2015 to 2019, the turnover rate of stocks on the NEEQ market dropped significantly, reaching 53.88%, 20.74%, 13.47%, 5.31%, and 6.00% respectively.

Third, the cost of listing on the NEEQ market is relatively high. In recent years, the regulation of the NEEQ market has become more stringent, leading to increased compliance costs for companies to get listed and companies that are already listed. In addition, companies listed on NEEQ need to pay management fees every year, and most of the NEEQ listed companies are small and medium-sized with relatively low profitability. Therefore, the high costs have forced those with poorer profitability to choose to leave.

Fourth, leading enterprises on the NEEQ market intend to get listed on other capital markets. Since 2018, most of the capital markets stared to intensify reforms. The HKEX allows enterprises with "weighted voting rights" and biotech enterprises that have not yet made profits to be listed. The SSE set up STAR (Sci-Tech innovation board) Market and launched a pilot program of the registration system. These moves have, to some extent, also attracted high-quality NEEQ listed companies to adjust

Table 1.2 Backdoor listing ZLCs

No	Before backdoor listing				After backdoor listing			
	Securities code	Listed time	Delisted time	Trade market	Securities code	Trade market	listing board	Listed time
1	430,007.OC	2006/9/7	2009/7/29	NEEQ	002,279.SZ	SZSE	SME board	2009/8/11
2	430,006.OC	2006/8/28	2009/9/30	NEEQ	300,016.SZ	SZSE	ChiNext Market	2009/10/30
3	430,001.OC	2006/1/23	2010/12/6	NEEQ	300,150.SZ	SZSE	ChiNext Market	2010/12/22
4	430,023.OC	2007/10/26	2011/4/20	NEEQ	300,213.SZ	SZSE	ChiNext Market	2011/5/5
5	430,008.OC	2006/8/30	2011/9/28	NEEQ	300,271.SZ	SZSE	ChiNext Market	2011/10/26
6	430,012.OC	2007/2/16	2012/5/10	NEEQ	300,318.SZ	SZSE	ChiNext Market	2012/5/23
7	430,045.OC	2009/2/18	2012/8/29	NEEQ	300,353.SZ	SZSE	ChiNext Market	2012/9/27
8	430,030.OC	2008/8/20	2014/1/9	NEEQ	300,370.SZ	SZSE	ChiNext Market	2014/1/23
9	430,049.OC	2009/2/18	2015/4/20	NEEQ	300,444.SZ	SZSE	ChiNext Market	2015/4/23
10	430,040.OC	2008/12/26	2015/4/22	NEEQ	300,445.SZ	SZSE	ChiNext Market	2015/4/24
11	430,018.OC	2007/9/19	2015/6/1	NEEQ	300,477.SZ	SZSE	ChiNext Market	2015/6/10
12	430,162.OC	2012/11/2	2017/8/24	NEEQ	/	/	/	/
13	831,008.OC	2014/8/29	2017/12/25	NEEQ	300,736.SZ	SZSE	ChiNext Market	2018/1/9
14	836,013.OC	2016/3/9	2018/1/19	NEEQ	/	/	/	/
15	834,801.OC	2016/1/5	2018/1/19	NEEQ	603,516.SH	SSE	Main Board	2018/2/2
16	834,116.OC	2015/11/10	2018/4/27	NEEQ	/	/	/	/
17	834,391.OC	2015/12/2	2018/10/10	NEEQ	688,078.SH	SSE	STAR Market	2019/12/30
18	870,077.OC	2016/12/6	2018/10/12	NEEQ	/	/	/	/
19	837,335.OC	2016/5/24	2018/12/6	NEEQ	/	/	/	/
20	839,570.OC	2016/11/8	2019/3/5	NEEQ	688,168.SH	SSE	STAR Market	2019/9/6
21	834,360.OC	2015/11/19	2019/4/1	NEEQ	688,277.SH	SSE	STAR Market	2020/7/7
22	836,201.OC	2016/4/8	2019/4/23	NEEQ	/	/	/	/

(continued)

Table 1.2 (continued)

No	Before backdoor listing				After backdoor listing			
	Securities code	Listed time	Delisted time	Trade market	Securities code	Trade market	listing board	Listed time
23	835,099.OC	2015/12/29	2019/5/22	NEEQ	/	/	/	/
24	834,291.OC	2015/11/26	2019/6/19	NEEQ	300,788.SZ	SZSE	ChiNext Market	2019/7/5
25	832,340.OC	2015/4/23	2019/7/8	NEEQ	603,613.SH	SSE	Main Board	2019/7/30
26	430,002.OC	2006/1/23	2019/8/1	NEEQ	603,927.SH	SSE	Main Board	2019/9/9
27	430,011.OC	2007/1/23	2019/9/30	NEEQ	300,803.SZ	SZSE	ChiNext Market	2019/11/18
28	430,581.OC	2014/1/24	2019/11/29	NEEQ	688,181.SH	SSE	STAR Market	2020/1/6

Remarks"/" means Wind has not made clear statistics yet
Source Wind, compiled by ZLCA

their strategies and participate in domestic and overseas capital markets. In 2019, a total of 7 former Zhongguancun NEEQ listed companies successfully got listed on the A-share market, and the number of successful backdoor listing this year was 3.5 times of 2018.

Fifth, some enterprises were forced to delist due to violations of regulations and failure to disclose annual reports or semi-annual reports within the prescribed time limit. To sum up, enterprises either take the initiative to delist for the high cost of listing and strategic adjustments, or are forced to delist due to violations of regulations. The leaving of a large number of NEEQ listed companies, to a certain extent, also helped solve the current difficulties such as the unsatisfactory enterprise quality and inactive transactions, accelerating survival of the fittest enterprises on the NEEQ market.

With the announcement of the China Securities Regulatory Commission (CSRC) on October 25th, 2019 to fully deepen the NEEQ reform, the process of NEEQ reform has further sped up. The reform is centered on facilitating connection of the multi-level capital market. In the overall NEEQ reform, the CSRC has promoted the optimization of bond issuance and financing system, the improvement of market classification and supporting systems, the establishment of a backdoor listing mechanism for listed enterprises, strengthening of supervision, foundation of market withdrawal mechanism and other reform measures. The NEEQ reform is not only conducive to the formation of a market-oriented mechanism of survival of the fittest, promoting the quality and efficiency of the NEEQ market, but also conducive to addressing the weakness of the multi-level capital market, and effectively serve SMEs and the private economy.

1.1.2 Companies Continuing Operations Achieved Increase in Revenue and Net Profit

Companies continuing operations have achieved increase in both revenue and net profit. In 2019, the total revenue and net profit of Zhongguancun NEEQ listed companies were 206.683 billion yuan and 8.137 billion yuan respectively, decreasing by 13.64 and 11.82% year on year. From 2018 to 2019, there were 1,035 Zhongguancun NEEQ listed companies continuing operations whose revenue and net profit in 2019 were 205.121 billion yuan and 8.753 billion yuan, with an increase of 6.53 and 21.95% year on year respectively (see Table 1.3). Based on the number of listed companies under investigation in this report, the decline in the total revenue and net profit of these companies in 2019 was mainly due to the decrease in the number of NEEQ listed companies rather than the performance of enterprises continuing operations. In 2019, despite the complex and severe external environment and the increasing downward pressure on economy, the performance of Zhongguancun NEEQ enterprises still maintained strong growth, indicating an improvement in the quality of Zhongguancun NEEQ listed companies.

In terms of average revenue and average net profit, companies on Innovation tier are three times as large as those on Base tier. The total revenues of Zhongguancun NEEQ listed companies on Innovation tier and those on Base tier were 47.772 billion yuan (23.11%) and 158.910 billion yuan (76.89%) respectively. At the meantime, the average revenue of companies on Innovation tier reached 531 million yuan while those on Base tier yielded an average revenue of 160 million yuan, with the former 3.31 times of the latter (see Table 1.4). Meanwhile, in 2019, the total net profits of Zhongguancun NEEQ listed companies on Innovation tier and those on Base tier

Table 1.3 Changes in revenue and net profit growth of Zhongguancun NEEQ listed companies in 2019

	Revenue		Net profit	
	Scale	Growth rate (%)	Scale	Growth rate (%)
Total	206.683	−13.64	8.137	−11.82
Continuing operations	205.121	6.53	8.753	21.95

Source Wind, compiled by ZLCA
Unit Billion yuan

Table 1.4 Classified distribution of revenue of Zhongguancun NEEQ listed companies in 2019

	Revenue/billion yuan	Percentage (%)	Average revenue/million yuan	Increase rate of average revenue (%)
Innovation	47.772	23.11	531	0.95
Base	158.91	76.89	160	0.63

Source Wind, compiled by ZLCA

were 1.929 billion yuan (23.71%) and 6.208 billion yuan (76.30%) respectively; the average net profit of the former was 21, 436, 500 yuan, and that of the latter was 6, 263, 900 yuan, with the former 3.42 times that of the latter.

Over 30% of the companies have achieved growth in both revenue and net profit, but the overall profitability still needs to be improved. 55.13% of the enterprises (596) achieved positive growth in revenue, and 51.16% of the enterprises (553) achieved positive growth in net profit. Among them, 35.15% of enterprises (390) achieved growth both in revenue growth and net profit. Separately, 35.62% of the enterprises (385) have achieved positive growth in revenue for two consecutive years, and 81 of them have achieved positive growth in revenue for two consecutive years with a compound growth rate not less than 50%; 47% of the enterprises (508) has been profitable for two consecutive years. Among the profitable enterprises, 20 have yielded an average net profit over 100 million yuan for two consecutive years, and 129 Zhongguancun NEEQ listed companies generated an average net profit of more than 20 million yuan.

1.1.3 NEEQ Listed Companies with Better Quality Need Stronger Financing Support

In 2019, the overall liquidity of the Zhongguancun NEEQ market continued to decline, but the quality of listed companies has improved compared with 2018, and the performance of leading companies in the capital market has shown signs of rebound. But judging from the angle of financing, the scale of corporate financing does not give rise to much optimism. The financing situation is especially unfavorable for private enterprises.

1.1.3.1 Stock Issuance

In 2019, both the number of stock issuances and financing on Zhongguancun NEEQ market continued to decline, but the average single financing scale rose against the trend. The number of stock issuance throughout the year was 79, a year-on-year decrease of 55.61%; the total financing scale was 4.099 billion yuan, a year-on-year decrease of 50.62%; the average single financing amount was 52 million yuan, a year-on-year increase of 15.30% (see Table 1.5).

1.1.3.2 Stock Trading volume

Zhongguancun NEEQ listed companies rebounded in the capital market. In 2019, a total of 535 Zhongguancun NEEQ listed companies participated in stock transactions, a year-on-year decrease of 23.18% compared with 659 in 2018. The annual

Table 1.5 Stock issuance on Zhongguancun NEEQ market

Year	Stock issuance		Financing scale		Average financing scale	
	Number of fiancing	Growth rate (%)	Scale (billion yuan)	Growth rate (%)	Scale (billion yuan)	Growth rate (%)
2015	436	373.91	33.123	304.98	0.076	−14.55
2016	511	17.20	34.3.6	3.57	0.067	−11.63
2017	419	−18.00	27.61	−19.52	0.066	−1.85
2018	186	−55.61	8.301	−69.93	0.045	−32.27
2019	79	−57.53	4.099	−50.62	0.052	15.30

Source NEEQ, compiled by ZLCA

stock trading volume was 1.695 billion shares and the turnover was 4.166 billion yuan. The trading volume and turnover were slightly lower than the 1.734 billion shares and 4.237 billion yuan in the same period in 2018; the average volume and average turnover were 3.241 million shares and 77.869 billion yuan; higher than the 2.631 million shares and 62.494 billion yuan in 2018, an increase of 23.17 and 24.60% year on year respectively; the annual average stock price was 2.46 yuan per share, basically the same as in 2018 (see Fig. 1.4). Therefore, it can be seen that the decreasing of of NEEQ listed companies participating in stock trading has not affected the NEEQ market trading conditions. Instead, it indicates a sign of recovery in the Zhongguancun NEEQ market.

The leading Zhongguancun NEEQ companies stock trading scale rose sharply. The number of companies with an annual turnover above 10 million shares was 34 with a total trading scale at 3.247 billion yuan. The number and scale have witnessed an increase of 25.93% and 40.32% respectively. In addition, the number and transactions of enterprises with annual turnover in other ranges have declined to varying degrees. Specifically, 11 enterprises achieved an annual trading scale of 5–10 million shares at 242 million yuan, down by 50.00 and 48.51% year on year; 72 enterprises

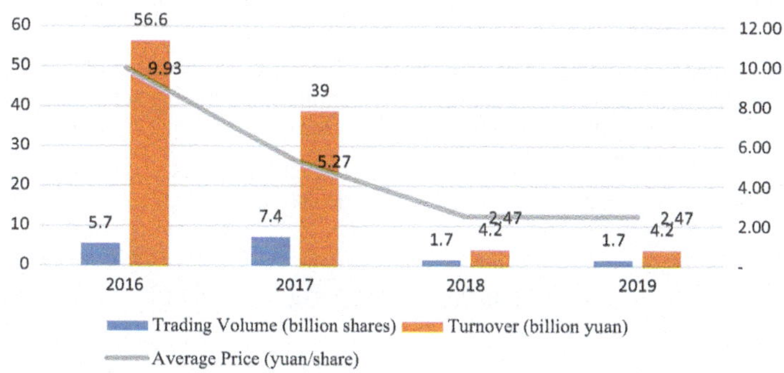

Fig. 1.4 2016–2019 annual trading volume and Turnover of Zhongguancun NEEQ listed companies. *Source* Wind, compiled by ZLCA

realized a trading volume of 1–5 million shares at 429 million yuan, a year-on-year decrease of 30.10% and 49.17% respectively; 418 enterprises had an annual transaction volume less than 1 million shares with the total turnover at 248 million yuan, a year-on-year decrease of 17.55 and 57.31%. Meanwhile, 546 Zhongguancun listed companies conducted no stock transaction in 2019, accounting for the majority of the total at 50.50% (see Table 1.6).

NEEQ enterprises that adopted market making trading performed better than those adopted call auction trading whether in transaction activity, turnover or average price. From the perspective of trading activity, only one of the 97 market-making enterprises had zero stock trading volume, while the number for enterprises with call auction trading were 439, accounting for 44.61% of the total (984). In terms of the turnover, market-making enterprises generated 2.532 billion yuan, accounting for 60.78%; the turnover of enterprises adopting call auction trading was 1.634 billion yuan, accounting for 39.22%. In terms of average price, market-making enterprises was 3.26 yuan per share, 1.83 times that of enterprises trading with call auction (see Table 1.7).

Enterprises on Innovation tier have obvious advantages in trading activity and average price, with less than 20% of the enterprises on this tier reaching more than

Table 1.6 Transactions of Zhongguancun NEEQ listed companies in 2019

	Numbers	Trading volume (million yuan)	Proportion of trading volume (%)
>= 0.1 billion shares	34	3247	77.94
5 million–0.1billion shares	11	242	5.81
1 million–5million shares	72	429	10.30
< 1 million shares	418	248	5.95
Total	535	4166	100.00

Source Wind, compiled by ZLCA

Table 1.7 Stock volume (Call Auction Trading and Market Making Trading) of Zhongguancun NEEQ companies in 2019

Trading mode	Enterprise number	Trading volume (million shares)	Proportion of trading volume (%)	Turnover (billion yuan)	Proportion of trading scale (%)	Average price (yuan/share)
Call auction trading	439	919	54.22	1.634	39.22	1.78
Market making trading	96	776	45.78	2.532	60.78	3.26
Total	535	1695	100.00	4.166	100.00	2.46

Source Wind, compiled by ZLCA

Table 1.8 Classified stock trading (Innovation and Base) volume of Zhongguancun NEEQ listed companies in 2019

	Innovation		Base	
	Numbers	Proportion (%)	Numbers	Proportion (%)
Enterprises	90	16.82	445	83.18
Trading volume (million shares)	717	42.30	978	57.70
Trading scale (million yuan)	2703	64.88	1463	35.12
Average price (yuan/share)	3.77		1.50	

Source Wind, compiled by ZLCA

60% of the total turnover. From the perspective of trading activity, all 90 enterprises on Innovation tier conducted stock transaction, while the number of enterprises on Base tier that conducted stock trading was 445, accounting for 44.90% of the Base tier enterprises. From the perspective of total trading volume and turnover, Innovation enterprises traded 717 million shares at 2.703 billion yuan, accounting for 42.30% and 64.88% respectively; Base enterprises traded 978 million shares at 1.463 billion yuan respectively, accounting for 57.70% and 35.12% respectively. From the perspective of average price, Innovation tier enterprises sold at 3.77 yuan/share, 2.5 times that of Base tier enterprises (see Table 1.8).

1.1.3.3 Financing Situation

Compared with other listed enterprises, NEEQ listed enterprises have relatively restricted financing channels due to smaller asset scale and unstable growth. At present, NEEQ listed enterprises can finance through bank credit, private placement, corporate bonds, market-maker system, preferred stock, and equity pledges. Due to the fact that data related to some financing methods cannot be obtained through public channels, this report only analyzes the financing situation of Zhongguancun NEEQ listed companies raising funds through private placement and corporate bonds in 2019.

(1) Equity private placement

Generally speaking, companies can use private placement to raise funds when applying for listing or after getting listed. In private placement, the company may apply for one confirmation for an issue in installments, which simplifies the stock issuance review process. The funds raised by private placement are reserved for

Table 1.9 Private placement of Zhongguancun NEEQ listed companies

Year	Number of companies	Number of private placements	Actual fundraising scale (billion yuan)	Volumes of private placement (million shares)	Average fundraising scale (ten thousands)
2015	283	376	30.962	5584	8235
2016	404	461	28.454	6175	6172
2017	342	377	22.723	2674	6027
2018	134	144	4.714	836	3274
2019	64	67	3.740	700	5582

Source Wind, compiled by ZLCA

specific purposes and under strict supervision. They could usually be used to supplement working capital, repay bank loans and other debts, and to serve other purposes including project financing, equity incentives, and asset purchasing.[5]

The number of enterprises obtain investments through private placement, the number of their private placements and the actual scale of funds being raised have all been at the lowest level in the past five years. But the average single fundraising scale has substantially increased after five consecutive declines. In 2019, a total of 537 NEEQ listed companies across the country completed 571 private placements, and the actual fundraising scale was 25.148 billion yuan, with an average of 44.04 million yuan in a single fundraising. Among them, 64 Zhongguancun NEEQ listed companies have implemented 67 private placements, and the actual scale of funds raised was 3.740 billion yuan, accounting for 14.87% of the total funds raised through private placements by NEEQ listed companies. The scale of single raised funds was 55.82 million yuan, 26.75% higher than the national average (see Table 1.9).

The private placement scale of more than 70% the listed enterprises was less than 50 million yuan. For the whole year of 2019, there were 10 private placements that raised over 100 million yuan and the actual fund raised was 2.548 billion yuan; there were 8 private placements raised funds between 50 and 100 million yuan, and the actual total fundraising scale was 588 million yuan; there were 23 private placements raised between 10 and 50 million yuan, and the actual total fundraising amounted to 491 million yuan; 26 private placements raised less than 10 million yuan with the actual total financing scale at 115 million yuan (see Table 1.10).

The average actual financing scale raised in a single fundraising by companies on Innovation tier was three times more than the scale raised by companies on Base tier. From the perspective of market classification, 14 enterprises on Innovation tier have implemented 14 private placements, and the actual fundraising scale was 1.635 billion yuan, 116.79 million yuan on average; 50 enterprises on Base tier completed

[5]The statistical standard of equity private placement in this book is the issuance date of private placement bonds. Equity private placement on NEEQ market refers to a sale of stock to pre-selected investors by companies applying for listing or having been listed on NEEQ market.

Table 1.10 Statistics on the Actual fundraising scale by private placement of Zhongguancun NEEQ listed companies in 2019

Fundraising range	Number of private placements		Actual fundraising scale	
	Numbers	Proportion (%)	Scale (million yuan)	Proportion (%)
>= 100million	10	14.93	2548	68.09
50 million–100 million	8	11.94	588	15.71
10 million–50 million	23	34.33	491	13.12
<10 million	26	38.81	115	3.07

Source Wind, compiled by ZLCA

53 private placements. The actual fundraising scale was 1.965 billion yuan, 37.08 million yuan on average (see Table 1.11).

(2) Issuance of Corporate Bonds

In terms of the issuance scale of corporate bonds, Zhongguancun NEEQ listed companies were the font-runners. In 2019, a total of 6 NEEQ listed companies conducted 9 corporate bond issuance raising 5.51 billion yuan. Among them, 3 Zhongguancun NEEQ listed companies conducted 9 corporate bond issuance with a fundraising scale at 5.34 billion yuan, accounting for 96.91% of the national NEEQ corporate bond issuance. To be specific, in 2019, the scale of corporate bonds issued by Zhongguancun NEEQ listed firms was 5.34 billion yuan, with bond issuance periods ranging from one to five years (mostly within a three-year period); the interest rates of most corporate bonds ranged from 6% to 6.5%. In addition, targeting at investors who opened qualified accounts of securities in registered enterprises (except investors prohibited by laws and regulations), China National Investment & Guaranty Corporation issued 5-year corporate bonds offline with a lower interest rate at 3.87%.

The scale of bond issuance by private companies accounted for less than 1%. From the perspective of company attributes, there are totally 5 corporate bond issuance

Table 1.11 Classified comparison of actual fundraising scale by private placement of Zhongguancun NEEQ listed companies in 2019

Classification	Number of companies	Number of private placements	Fund volume (million shares)	Actual fundraising scale (million yuan)	Average (ten thousands)
Innovation	14	14	89	1965	11,679
Base	50	53	503	1965	3708

Source Wind, compiled by ZLCA

conducted by 2 central SOEs (state-owned enterprises), with a total issuance of 5.3 billion yuan, accounting for 99.25%. Among them, China Kangfu International Leasing CO., LTD. conducted four issuance at 2.8 billion yuan; China National Investment and Guaranty Corporation conducted one issuance with a total of 2.5 billion yuan. The issuance of corporate bonds by private companies was only 40 million yuan, accounting for 0.75% of the total bond issuance volume.

1.1.4 Companies in Majority Achieved Positive Growth in R&D Investment, and PCT Patent Applications Continued to Rise in Number

Over 50% of Zhongguancun NEEQ listed companies have achieved positive growth in R&D expenses. In 2019, a total of 862 (accounting for 79.74%) disclosed their R&D expenses which amounted to 8.45 billion yuan, a year-on-year decrease of 15.75%. From 2018 to 2019, 806 companies that continuously disclosed R&D expenses invested a total of 8.244 billion yuan in R&D, which was basically the same as in 2018. This shows that the total R&D expenses of Zhongguancun NEEQ listed companies declined because of the decrease in the number of listed enterprises. A detailed analysis of the growth of R&D expenses of the 806 enterprises indicates that 452 enterprises (56.08%) have achieved positive growth in R&D expenses, and 354 enterprises (43.92%) have experienced a year-on-year decline in R&D expenses.

The R&D intensity of leading Zhongguancun NEEQ listed companies continued to increase. In 2019, the average R&D intensity of Zhongguancun NEEQ listed companies was 4.70%, which was higher than the average of 4.15% of all NEEQ listed companies and 2.19% in the country in 2019. International experience has shown that R&D intensity over 10% indicates a relatively high level of R&D investment, and companies as such are generally considered to have sufficient R&D competitive advantages. According to this standard, the R&D intensity of Zhongguancun NEEQ listed companies is rather satisfactory. In 2019, 368 enterprises were considered to have high R&D intensity, accounting for 42.7%, an increase of 2.7 percentage points year on year. It shows that the downward macroeconomic fundamentals has little impact on enterprises with high R&D intensity. The leading Zhongguancun NEEQ listed companies have improved their core competitiveness by increasing R&D investment against the economic trend.

From the perspective of patent application and patent authorization, in 2019, Zhongguancun NEEQ listed companies made a breakthrough in the number of PCT applications, reaching 25. In comparison with 17 applications in 2018, it is an increase of 47%. The data shows that Zhongguancun NEEQ listed companies has attached great importance to patent layout in overseas markets. In 2019, the number of patent applications on the Zhongguancun NEEQ market was 2,362. Among them, there were 914 invention patents, 1,158 utility model patents and 290 design patents, accounting for 38.70, 49.03 and 12.27% of the total patent applications respectively.

The number of patent authorization was 1,450, lower than 2,144 in 2019. Among them, the number of invention patents, utility model patents, and design patents were 305, 918, and 227 respectively, accounting for 21.03%, 63.31%, and 15.66% respectively. Comparing with the proportions of different types of patent applications, it is found that the proportion of authorized invention patents of Zhongguancun NEEQ listed companies is relatively low. This may be related to the long review period of invention patents and the low approval rate.

1.1.5 Obvious Regional Competitive Advantage Over Other Regions

This report conducts a comparative analysis of the top five regions with the highest number of NEEQ listed companies. To grasp a full knowledge of the growth of listed companies in the five regions, this report investigates such core indicators as total market value, revenue, net profit, total assets, financing, taxation, and employee number. The statistics shows that in 2019 when the overall macroeconomic continue to recess, the overall performance of the NEEQ listed companies in major regions has shown a downward trend. However, performance decline is mainly caused by the delisting of enterprises as opposed to a decrease in the performance of enterprise continuing operations. Moreover, Z-Park still has a strong competitive advantage compared with other regions.

From the perspective of total market value, NEEQ listed companies of Z-Park ranked the first in China. At the end of 2019, the total market value of Zhongguancun NEEQ listed companies was 340.05 billion yuan, followed by other regions of Guangdong, Jiangsu, Zhejiang, and Shanghai at 316.388 billion yuan, 217.223 billion yuan, 184.549 billion yuan, and 216.884 billion yuan respectively. In terms of the market value of leading enterprises, both Zhongguancun and Guangdong Province got three NEEQ listed enterprises valuing more than 10 billion yuan; Both Zhejiang and Shanghai got one enterprise valuing over 10 billion yuan; all NEEQ listed enterprises in the Jiangsu Province valued below 10 billion yuan.

From the perspective of total assets, Zhongguancun NEEQ listed companies lead the way around the country, but the risk of decline remains. In 2019, the total assets of Zhongguancun NEEQ enterprises ranked the first at 368.225 billion yuan, followed by NEEQ listed enterprises in Jiangsu with a total of 314.329 billion yuan; NEEQ listed companies in Guangdong, Zhejiang, and Shanghai had total assets of 285.542 billion yuan and 218.42 billion yuan and 182.220 billion yuan respectively. In 2019, the total assets of NEEQ listed companies in most regions have declined to varying degrees, except those in Zhejiang Province which realized a slight increase of 0.95%.

From the angle of revenue, Z-Park listed NEEQ companies were in the middle. In 2019, Zhongguancun NEEQ listed companies harvested a total of 206.683 billion yuan, higher than the total revenue of 181.638 billion yuan yielded by NEEQ listed companies in Jiangsu, but lower than the number in the three regions of Shanghai,

Zhejiang and Guangdong. Judging from the total revenue of the NEEQ listed enterprises continuing operations for two consecutive years of 2018 and 2019, the total revenue in the above five regions has increased in 2019. Among them, NEEQ listed companies continuing operations in both Zhongguancun and Guangdong increased by 6%, lower than those in Shanghai, Jiangsu and Zhejiang.

As to net profit, Zhongguancun NEEQ listed companies in 2019 harvested 8.137 billion yuan, which was lower than the number of Jiangsu (9.128 billion yuan) and Zhejiang (8.998 billion yuan), but higher than Guangdong (6.279 billion yuan) and Shanghai (3.017 billion yuan). From the perspective of average and per capita net profit, Zhongguancun NEEQ listed companies showed advantages in the above five major regions. The average net profit was 7.5273 million yuan, higher than that of Guangdong and Shanghai NEEQ listed companies; the per capita net profit of Zhonguancun listed companies was 47,600 yuan/person, second only to those in Zhejiang Province which was 50,100 yuan/person. Judging from the net profit of NEEQ listed companies continuing operations for two consecutive years from 2018 to 2019, Z-Park NEEQ listed companies are higher than those in other regions. At the meantime, NEEQ listed companies continuing operations enjoyed a growth rate in net profit as high as 21.95%.

In financing, the scale of fund raised by Z-Park NEEQ listed companies through stock issuance and bond issuance was far ahead. In 2019, the scale of funds raised by NEEQ listed enterprises nationwide through stock issuance and bond issuance has decreased significantly. In line with the national trend, the scale of Zhongguancun NEEQ listed companies also declined, but the amount of funds raised was still far ahead. In 2019, Z-Park NEEQ listed companies raised a total of 9.439 billion yuan through the above two methods. Among them, the scale of funds raised through stock issuance was 4.099 billion yuan, and funds raised through bond issuance was 5.34 billion yuan.

1.1.6 Accounts Receivable of Enterprises Continuing Operations Under Pressure with Longer Accounts Receivable Days

Accounts receivable (AR) is an important current asset for an enterprise, but with high-risk. The collection of AR will directly affect the cash flow and working capital of a company. Under the circumstances of unsound AR management mechanism, large the scale of AR, or long days sales outstanding (DSO), the difficulty of payment collection and the possibility of bad debts would increase. In the most severe cases, it could become the final straw. In 2019, the total scale of AR for Zhongguancun NEEQ listed companies was 55.724 billion yuan, a year-on-year decrease of 6.60%; the average scale was 51.55 million yuan, an increase of 13.13% year on year; the DSO were 95.17 days, increased by 7 days compared to 2018. From 2018 to 2019, the AR of 1,035 enterprises with continuing operations for two consecutive years was 54.672 billion yuan, an increase of 7.81% from 50.713 billion yuan in 2018, an increase rate that is 0.58 percentage points higher than 2018. This set of data showed that in 2019, the total AR of Zhongguancun NEEQ listed companies decreased mainly due to the delisting of listed enterprises, while the AR scale and the DSO of listed enterprises continued to rise. The overall economic downturn increased the pressure to collect payment, especially for small and medium-sized downstream enterprises. As a result of these factors, it was more difficult for enterprises to collect payments.

Researches on companies find that the major reasons for the continuous increase in the AR of Zhongguancun NEEQ listed companies are as follows: (1) In order to expand the market share, companies choose to sell on credit or extend the payment days; (2) Some companies face barriers in collecting payments due to government budget reduction, cumbersome review process, and long payment cycle; (3) The central and local SOEs and other core companies have difficulty in rights confirmation, making it difficult for upstream companies to collect payments, which in turn leads to a prominent problem of "triangle debt" between upstream and downstream companies in the industrial chain.

1.1.7 Company Tax Burden Reduced, But There is Room for Further Reduction

From 2018 to 2019, the total profit of Zhongguancun companies continuing operations in 2019 increased by 21.96% while the income tax was reduced by 6.06%; the actual tax burden of companies continuing operations was 7.099 billion yuan, decreased by 5.16% from the 7.485 billion yuan in 2018. Data shows that under the national environment of cutting taxes and fees, the actual tax burden of SMEs is indeed much lighter than in previous years. However, the actual tax burden of Zhongguancun NEEQ listed enterprises now is still relatively high, and there is room of further reduction. In 2019, the actual tax burden of Zhongguancun NEEQ listed companies accounted for 3.45% of the revenue, higher than the 3.38, 2.73, 2.03, and 1.54% in Jiangsu, Guangdong, Zhejiang, and Shanghai regions, indicating the actual tax burden of Zhongguancun NEEQ listed companies could continue to be reduced.

1.2 Suggestions for the Development of Zhongguancun NEEQ Listed Companies

1.2.1 Innovate Financing Methods and Broaden Financing Channels

Around the country, the downward pressure on the economy is increasing. For Zhongguancun NEEQ listed companies, the slow growth of income and increase in accounts receivable year by year, coupled with barriers of financing through bank credit, bond issuance, private placement and equity pledge, have made it difficult for enterprises to obtain stable cash inflows. On the contrary, corporate cash outflows have indeed increased year by year, which has led to difficulties in cash flow turnover for some enterprises who urgently need to expand financing channels.

Data shows that the cash inflow from operating activities of continuing operations[6] in 2019 was 243.484 billion yuan, a year-on-year decrease of 0.59%; the cash outflow was 235.032 billion yuan, increased by 2.63% compared with 2018; the net cash flow from operating activities was 8.452 billion yuan, decreased by 46.93% at 7.474 billion yuan year on year. At the same time, from 2018 to 2019, the accounts receivable of the 1,035 Zhongguancun NEEQ listed companies that continued to operate for two consecutive years was 54.672 billion yuan, an increase of 7.81% from the 50.713 billion yuan in 2018. The increase rate was 0.58 percentage points higher than 2018. In terms of financing, in 2019, Zhongguancun NEEQ listed companies actually raised 3.740 billion yuan through private placement, a year-on-year decrease of 21.66%. Among the 5.34 billion yuan raised through bonds, private enterprises only accounted for 40 million yuan. This set of data indicates the cash flow pressure on NEEQ listed companies.

In response to the cash flow problems of Zhongguancun NEEQ listed companies, this report proposes policy recommendations from two aspects: expanding the direct financing and indirect financing channels.

To broaden direct financing channels, there are two suggestions. **First is to seize the opportunity of the NEEQ reform to increase the proportion of direct financing for NEEQ listed companies**. It is recommended that NEEQ should avail itself of the reform opportunity to establish the market mechanism for the survival of the fittest, improve the overall quality of the NEEQ enterprises, so as to increase the liquidity of the NEEQ market and the direct financing ratio and financing efficiency. In so doing, the NEEQ could practically serve innovative, entrepreneurial and small, medium and micro enterprises that are not full-fledged. **The second suggestion is to establish a special fund to support the development of enterprises that are in temporary difficulty**. It is recommended that the Beijing Municipal Government set up a special policy fund to bail out Zhongguancun NEEQ listed companies that are technology-intensive with great market potential when they are face temporary trouble in operation.

In terms of expanding indirect financing channels, **first, it is recommended to subsidize the innovative financial products of social organizations in an appropriate proportion**. After years of continuous research, Zhongguancun Listed Companies Association has sorted out the sticking point causing difficulties in AR collection for some enterprises, and has actively launched an innovative financing service program in cooperation with financial institutions. The program is well accepted by companies. As the financing product is not subsidized by the government, many enterprises have expressed their hope to include this project into the

[6]There are a total of 1,035 Zhongguancun NEEQ enterprises with continuing operations from 2018 to 2019.

scope of subsidies, which could bring tangible benefits to companies. **Second, the government should encourage enterprises to innovate financing models**. Traditional financing methods focus on credit rating, assets, and cash flow of corporate entities. However, the scale of the NEEQ listed companies is generally small with limited revenue. Therefore, it is difficult for these companies to finance through bank loans and secondary market stock pledges. The government should encourage enterprises to innovate financing models, such as by changing their business models to obtain untraditional financing channels. **The third is to promote the implementation of inclusive financing**. It is recommended that relevant government departments actively promote the practical implementation of inclusive financial support measures to clear up the difficulties to accessible financing for small, medium and micro enterprises.

1.2.2 Increase the Quality of R&D Output to Improve Core Competitiveness

Continuous R&D investment is the propeller for enterprises to quickly occupy the market and obtain excess profits. Therefore, focusing on R&D and improving the quality of R&D output is the secret weapon to improve core competitiveness. However, it can be seen from the data on the innovation capability of the NEEQ listed companies in 2019 that the quality of innovation output needs to be further improved. From the perspective of patent applications and patent authorizations, in 2019, the number of patent applications on the Zhongguancun NEEQ market was 2,362, of which the numbers of invention patents, utility model patents and design patents were 914, 1,158 and 290, respectively, accounting for 38.70, 49.03, and 12.27% of the total; the number of patent authorizations was 1,450, lower than 2,144 in 2019. Among them, the number of invention patents, utility model patents, and design patents were 305, 918, and 227, accounting for 21.03%, 63.31%, and 15.66% respectively. From the proportions of different types of patent applications, it is found that the proportion of authorized invention patents of Zhongguancun NEEQ listed companies is relatively low. This may be related to the long examination period of invention patents and the low approval rate.

The suggestions are proposed to increase the approval rate of invention patents of Zhongguancun NEEQ listed companies. **First is to improve the efficiency of patent application examination**. It is recommended that relevant government departments of Beijing reduce unnecessary formalities for applicants, facilitate the patent examination process and shorten the period of invention patent authorization. **The second is to strengthen the integration of production, education and research**. It is suggested that Zhongguancun NEEQ listed companies rely on external resources such as universities, scientific research institutions and technical service institutions to further enhance their innovation capabilities and improve the quality of their innovation output.

1.2.3 Step Up Efforts to Reduce Taxes and Fees and Ease the Burden on Enterprises

In 2019, the actual tax burden of Zhongguancun NEEQ listed companies continuing operations was 7.099 billion yuan, decreased by 5.16% from the 7.485 billion yuan in 2018. This is mainly due to the efforts to cut taxes and fees in recent years. In particular, on April 7th, 2019, The General Office of the CPC Central Committee and the General Office of the State Council printed and issued the "Guiding Opinions on Promoting the Healthy Development of Small and Medium-Sized Enterprises" (hereinafter referred to as the Guiding Opinions). The Guiding Opinions put forward the following measures: clean up and standardize business-related charges, and accelerate zero-charge on local business-related administrative fees; promote substantial tax cuts including value-added tax, and provide tax exemptions for micro and small firms as well as technology startups; and lower the nominal rates for social security contributions in accordance with actual conditions to support SOEs to absorb employment. However, the actual tax burden of enterprises accounted for 3.45% of the revenue, which was higher than the 3.38, 2.73, 2.03, and 1.54% in Jiangsu, Guangdong, Zhejiang, and Shanghai regions, indicating that the actual tax burden of Zhongguancun NEEQ listed companies could be further eased.

In view of the excessive tax burden of Zhongguancun NEEQ listed companies, it is recommended that relevant government departments provide preferential taxes and fees to strategic emerging industries in energy conservation and environmental protection, new generation of information technology, biology, high-end equipment manufacturing, new energy, new materials, and new energy vehicles. Preferential taxes and fees will foster the development and growth of such industries and form an industry cluster with strong combined effect in Z-Park.

Chapter 2
Profitability Analysis of Zhongguancun NEEQ Listed Companies in 2019

Abstract This report analyzes the profitability of Zhongguancun NEEQ enterprises, and analyzes the overall operation status from six aspects of operating income, gross profit, net profit, return on total assets, return on equity, and period expenses to fully reflect corporate profitability. According to the report conclusion, the operating income, gross profit, net profit, and gross profit margin of Zhongguancun NEEQ enterprises all declined to a certain extent, and the net profit margin, return on total assets, and return on net assets rose to some extent. In view of the continuous operation of Zhongguancun NEEQ enterprises in 2018–2019, the business income, gross profit and net profit have all increased, indicating that the profitability of continuous operation enterprises has been significantly improved. In addition, the sales capacity and management efficiency of Zhongguancun NEEQ enterprises have increased year by year.

Keyword Operating income · Net profit · Profitability

2.1 Revenue

2.1.1 Overall Revenue

The number of Zhongguancun NEEQ Listed Companies has decreased, but the overall quality has steadily improved. In 2019, the total revenue of Zhongguancun NEEQ Listed Companies was 206.683 billion yuan, a year-on-year decrease of 13.64%; the average revenue of them was 191 million yuan, a year-on-year decrease of 2.55% (see Fig. 2.1). From 2018 to 2019, there were 1035 Zhongguancun NEEQ Listed Companies that continued to operate for two consecutive years. The revenue of companies in continuous operation in 2019 was 205.121 billion yuan, and in 2018 was 192.556 billion yuan, representing an increase of 6.53% year-on-year. Based on the number of Zhongguancun NEEQ Listed Companies included in the report in

Zhongguancun Listed Companies Association, *The Growth Report of Zhongguancun NEEQ Listed Companies (2020)*, Current Chinese Economic Report Series,
https://doi.org/10.1007/978-981-33-6819-4_2

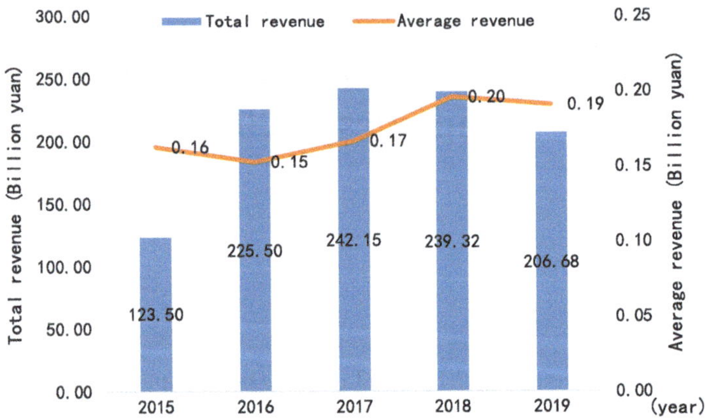

Fig. 2.1 Changes in total revenue and average revenue of Zhongguancun NEEQ listed companies from 2015 to 2019. *Source* Wind, compiled by ZLCA.

Table 2.1 Distribution of revenue of Zhongguancun NEEQ listed companies in 2019

Classification	Revenue/Billion yuan	Proportion of revenue (%)	Average revenue/Billion yuan	Average revenue growth (%)
Innovation	47.77	23.11	0.53	0.95
Base	158.91	76.89	0.16	0.63

Source Wind, compiled by ZLCA

2019,[1] the decline in the total revenue of Zhongguancun NEEQ Listed Companies in 2019 was mainly due to the decrease in the number of companies listed on the NEEQ, rather than the declining performance of companies in continuous operation, which instead have maintained robust growth.

The total revenue of Zhongguancun NEEQ Listed Companies on innovation tier was 47.772 billion yuan, accounting for 23.11%; the total revenue of those on base tier was 158.910 billion yuan, accounting for 76.89%. The average revenue of Zhongguancun NEEQ Listed Companies on innovation tier was 531 million yuan, a year-on-year increase of 0.95%; the average revenue of those on base tier was 160 million yuan, a year-on-year increase of 0.63% (see Table 2.1). It can be seen from the data that the average revenue of Zhongguancun NEEQ Listed Companies on innovation tier was 3.31 times that of companies on base tier, and the average revenue of companies on innovation tier increased more rapidly.

In 2019, there were 596 Zhongguancun NEEQ Listed Companies that achieved positive growth in revenue, accounting for 55.13% of the total. Among them, 60 were on innovation tier (accounting for 66.67% of all companies on innovation tier), and

[1] At the end of 2018, the number of Zhongguancun NEEQ Listed Companies included in the report was 1224; at the end of 2019, the number was 1081, a year-on-year decrease of 143, namely 12%.

536 were on base tier (accounting for 54.09% of all companies on base tier). Of those 596 companies, 385 achieved positive growth in revenue for two consecutive years from 2018 to 2019, accounting for 35.62% of the total number of Zhongguancun NEEQ Listed Companies. Among them, 37 were on innovation tier (accounting for 41.11% of all companies on innovation tier), 348 were on base tier (accounting for 35.11% of all companies on base tier). From 2018 to 2019, there were 81 companies with continuous growth in revenue and a compound growth rate of not less than 50%. These Zhongguancun NEEQ Listed Companies are in a stage of high growth. The top 30 Zhongguancun NEEQ listed companies by revenue growth in 2019 (see Table 2.2).

2.1.2 Size Distribution

Of all Zhongguancun NEEQ Listed Companies, there were 27 with a revenue of 1 billion yuan or more, which accounted for 2.50%, and the aggregate revenue of them reached 95.559 billion yuan, taking up 46.23% of the total; there were 39 with a revenue of 500 million to 1 billion yuan, accounting for 3.61%, and the aggregate revenue of them was 26.491 billion yuan, taking up 12.81% of the total; there were 120 with a revenue of 200 million yuan to 500 million yuan, accounting for 11.10%, and the aggregate revenue of them was 37.155 billion yuan, taking up 17.98% of the total; there were 163 with a revenue of 100 million yuan to 200 million yuan, accounting for 15.08%, and the aggregate revenue of them was 22.772 billion yuan, taking up 11.02% of the total; there were 200 with a revenue of 50 million yuan to 100 million yuan, accounting for 18.50%, and the aggregate revenue of them was 14.336 billion yuan, taking up 6.94% of the total; there were 532 with a revenue of less than 50 million yuan, accounting for 49.21%, and the aggregate revenue of them was 10.380 billion yuan, taking up 5.02% of the total. The size distribution shows a trend of polarization: nearly 50% of the total revenue was generated by large-scale companies (with a revenue of 1 billion yuan and above) that only accounted for 2.50%, while the revenue of the nearly 50% companies with a revenue of less than 50 million yuan only accounted for 5%. The revenue of most companies was at the middle and low level, which suggests that the Zhongguancun NEEQ Listed Companies are mainly in small and medium sizes (see Fig. 2.2).

A majority of Zhongguancun NEEQ Listed Companies on base tier reported a revenue of less than 50 million yuan, namely 522 companies, accounting for 52.67% of all. The numbers of companies with a revenue in the four ranges of 100 million yuan–200 million yuan, 200 million yuan–500 million yuan, 500 million yuan–1,000 million yuan, 1 billion yuan and above were on the decrease to138, 103, 26 and 21, accounting for 13.93%, 10.39%, 2.62% and 2.12% respectively. The revenue of companies on innovation tier was mainly in the three ranges of 50 million yuan–100 million yuan, 100 million yuan–200 million yuan and 200 million yuan–500 million yuan. The numbers of them were 19, 25 and 17, accounting for 21.11%, 27.78% and 18.89% respectively; There were 10 companies with a revenue of below 50 million yuan, accounting for 11.11%; 13 with a revenue of 500 million to 1

Table 2.2 The top 30 Zhongguancun NEEQ listed companies by revenue growth in 2019

Serial number	Code	Name	Tier	Industry	2018 revenue (thousand yuan)	2019 revenue (thousand yuan)	Year-on-year growth rate (%)
1	837,852.OC	Bond club	Base	Leasing and Business services	75.50	37,442.00	49,510.69
2	833,575.OC	Health guard	Base	Manufacturing	132.20	24,689.50	18,582.48
3	430,079.OC	Beijing anpeng	Base	Financials	75.60	3809.30	4942.01
4	872,434.OC	Mingyida	Base	Information transmission, Software and IT services	2699.30	116,153.10	4203.04
5	836,020.OC	Dawei	Base	Information transmission, Software and IT services	2382.70	52,710.10	2112.21
6	430,185.OC	Huaxin data	Base	Manufacturing	21.60	471.70	2085.16
7	430,189.OC	Modian	Base	Information transmission, Software and IT services	2041.70	17,956.00	779.46
8	836,008.OC	Moreget	Base	Manufacturing	14,824.80	119,569.90	706.55
9	835,014.OC	Space parking	Base	Leasing and Business services	211.50	1679.10	693.75

(continued)

Table 2.2 (continued)

Serial number	Code	Name	Tier	Industry	2018 revenue (thousand yuan)	2019 revenue (thousand yuan)	Year-on-year growth rate (%)
10	871,892.OC	Civigas	Base	Information transmission, Software and IT services	432.30	3261.90	654.6
11	830,871.OC	ADS	Base	Manufacturing	530.90	3883.50	631.47
12	837,317.OC	North point entertainment	Base	Culture, Sports and Entertainment	719.90	4379.20	508.34
13	834,465.OC	Sinoscience	Base	Information transmission, Software and IT services	17,332.90	102,742.00	492.76
14	872,287.OC	BCEG Advabced construction materials	Base	Manufacturing	73,882.00	431,352.60	483.84
15	430,167.OC	Sleton	Base	Manufacturing	4063.50	20,731.70	410.2
16	835,282.OC	Malon	Base	Culture, Sports and Entertainment	4001.00	18,587.80	364.58
17	873,069.OC	Tianlang zhida	Base	Scientific research and technology service	27,644.80	121,045.10	337.86
18	430,361.OC	Banggoo	Base	Manufacturing	966.00	4226.70	337.53

(continued)

Table 2.2 (continued)

Serial number	Code	Name	Tier	Industry	2018 revenue (thousand yuan)	2019 revenue (thousand yuan)	Year-on-year growth rate (%)
19	836,098.OC	Sinovinetechnology	Base	Information transmission, Software and IT services	3711.60	15,018.10	304.63
20	871,174.OC	Zhengrong networks	Base	Manufacturing	2434.60	9481.60	289.45
21	430,341.OC	Centran	Base	Information transmission, Software and IT services	7349.80	28,561.10	288.6
22	834,819.OC	Marr bio-pharmaceutical	Base	Manufacturing	21,683.30	83,069.60	283.1
23	835,101.OC	Smart-snapshot	Base	Information transmission, Software and IT services	832.70	3111.20	273.62
24	873,323.OC	Swgardens	Base	Construction	8202.10	30,408.10	270.73
25	838,983.OC	Beijing power communication equipment	Base	Manufacturing	2770.40	10,056.10	262.98
26	837,479.OC	Tianhe yuanchuang	Base	Information transmission, Software and IT services	4930.50	17,407.10	253.05

(continued)

Table 2.2 (continued)

Serial number	Code	Name	Tier	Industry	2018 revenue (thousand yuan)	2019 revenue (thousand yuan)	Year-on-year growth rate (%)
27	871,169.OC	Lanyun.net	Base	Information transmission, Software and IT services	58,401.90	183,866.70	214.83
28	430,170.OC	Jinyitong sci.&tech.beijing	Base	Manufacturing	7257.50	22,800.10	214.16
29	836,468.OC	Prova	Base	Manufacturing	8959.90	27,252.60	204.16
30	834,084.OC	Juneng dingli	Innovation	Scientific research and technology service	49,597.20	150,725.80	203.9

Source Wind, compiled by ZLCA

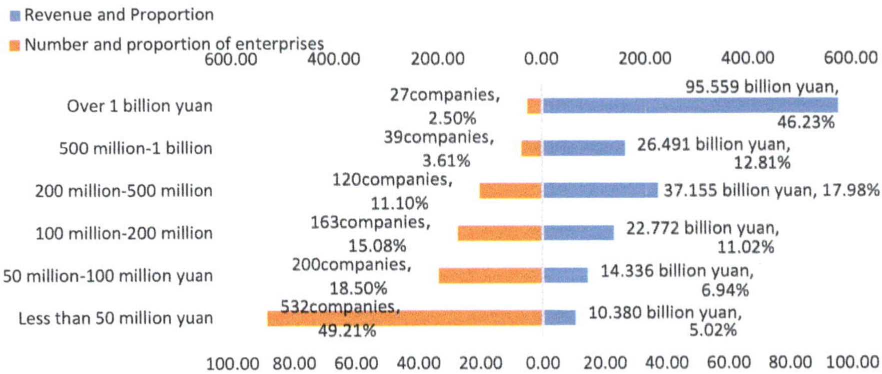

Fig. 2.2 Distribution of revenue of Zhongguancun NEEQ listed companies in 2019. *Source* Wind, compiled by ZLCA

billion yuan, accounting for 14.44%; 6 with a revenue of 1 billion yuan or more, accounting for 6.67% (see Table 2.3). This group of data shows: on the one hand, compared with the distribution of revenue of companies on base tier, that of those on innovation tier is more even, and the difference between the proportions of companies in different revenue scale ranges is quite small; On the other hand, there are quite

Table 2.3 Distribution of revenue of Zhongguancun NEEQ listed companies on innovation and base tiers in 2019

Scale	Total		Base		Innovation	
	Quantity	Proportion (%)	Quantity	Proportion (%)	Quantity	Proportion (%)
Over 1 billion yuan	27	2.50	21	2.12	6	6.67
500 million–1 billion	39	3.61	26	2.62	13	14.44
200 million–500 million	120	11.10	103	10.39	17	18.89
100 million–200 million	163	15.08	138	13.93	25	27.78
50 million–100 million yuan	200	18.50	181	18.26	19	21.11
Below 50 million yuan	532	49.21	522	52.67	10	11.11

Source Wind, compiled by ZLCA

many Zhongguancun NEEQ Listed Companies on base tier that reported a high revenue.

2.2 Gross Profit

In 2019, the total gross profit of Zhongguancun NEEQ Listed Companies reached 46.63 billion yuan, a year-on-year decrease of 18.82%; the average gross profit was 43 million yuan (see Fig. 2.3), a year-on-year decrease of 8.51%; the average gross margin was 22.56%, a year-on-year decrease of 5.99%. The total gross profit of 1,035 NEEQ companies that continued to operate in 2018 and 2019 was 46.398 billion yuan, a year-on-year increase of 5.26%; the average gross margin of companies in continuous operation was 22.62%, which is even with 22.89% in 2018. Data show that the decline in gross profit in 2019 is mainly due to the increase in the number of delisted companies in 2019, and the total gross profit of 1,035 companies that continued to operate increased on the whole.

The total gross profit of Zhongguancun NEEQ Listed Companies on innovation tier was 10.259 billion yuan, accounting for 22%, and the average gross profit was 114 million yuan; On base tier, the total gross profit was 36.372 billion yuan, accounting for 78.00%, and the average gross profit was 37 million yuan. The average gross profit of the innovation tier was 3.08 times that of the base tier. The gross margin of Zhongguancun NEEQ Listed Companies on innovation tier was 23%, and that of the base tier was 21%. The innovation tier was slightly higher than the base tier. On the whole, the gross profit of the innovation tier was higher than that of the base tier.

There were 584 Zhongguancun NEEQ Listed Companies that achieved positive growth in gross profit in 2019, accounting for 56.43% of all companies in continuous

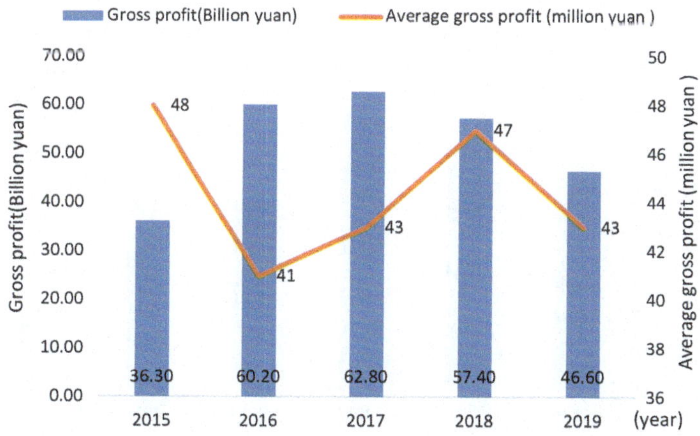

Fig. 2.3 Changes in gross profit and average gross profit of Zhongguancun NEEQ listed companies from 2015 to 2019. *Source* Wind, compiled by ZLCA

operation. More than half of Zhongguancun NEEQ Listed Companies achieved positive growth in gross profit. Among them, 71 Zhongguancun NEEQ Listed companies reported an increase in gross profit of more than 100%, and 7 more than 1,000%. On innovation tier, 63.33% (57) reported gross profit growth, while on base tier, the number was 53.08% (526).

2.3 Net Profit

2.3.1 Overall Net Profit

In 2019, the total net profit of Zhongguancun NEEQ Listed Companies was 8.137 billion yuan (see Fig. 2.4), a year-on-year decrease of 11.82%; the average net profit was 7.5271 million yuan, a year-on-year decrease of 0.17%. The net profit of the 1,035 Zhongguancun NEEQ Listed Companies that continued to operate in 2018 and 2019 reached 8.753 billion yuan, a year-on-year increase of 21.95%. Among them, 557 companies achieved positive growth in net profit, accounting for 53.82% of all companies in continuous operation. Data analysis shows that the decrease in the net profit of Zhongguancun NEEQ Listed Companies is mainly because too many companies have been delisted, and companies in continuous operation displayed good profitability.

The average net profit of companies on innovation tier in 2019 was higher than that of those on base tier. In 2019, the overall net profit of Zhongguancun NEEQ Listed Companies on base tier was 6.208 billion yuan, accounting for 76.30%, and the average net profit was 6.263 million yuan; the overall net profit of companies on innovation tier was 1.929 billion yuan, accounting for 23.71%, and the average

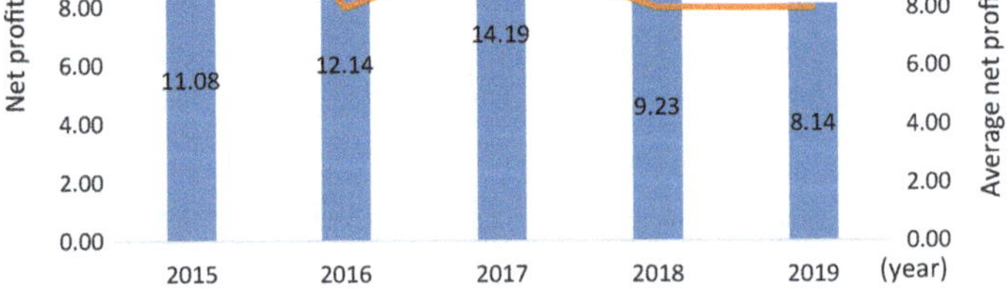

Fig. 2.4 Changes in total net profit and average net profit of zhongguancun NEEQ listed companies from 2015 to 2019. *Source* Wind, compiled by ZLCA

net profit was 21,436,500 yuan. The average net profit of Zhongguancun NEEQ Companies on innovation tier is much higher than that of those on base tier.

2.3.2 *Analysis of the Overall Profit and Loss of Zhongguancun NEEQ Listed Companies in 2019*

According to the overall profit and loss of Zhongguancun NEEQ Listed Companies from 2015 to 2019, the proportion of profitable companies has declined year by year. In 2019, there were 626 profitable companies, accounting for 57.91%; the number of non-profitable companies was 455, accounting for 42.09%. Overall, the profitability of Zhongguancun NEEQ Listed Companies needs to be improved (see Table 2.4).

In 2019, among Zhongguancun NEEQ Listed Companies on innovation tier, 60 (accounting for 66.67%) were profitable, 30 were operating in the red (accounting for 33.33%); on base tier, 566 companies were profitable (accounting for 57.11%), 425 were loss-making (42.89%) (see Table 2.5). In 2019, the profitability of Zhongguancun NEEQ Listed Companies on innovation tier was better.

There were 508 Zhongguancun NEEQ listed innovation Companies that have been profitable for two consecutive years in 2018 and 2019, accounting for 47.00% of the total. Among them, 56 were on innovation tier (62.22%), 453 on base tier (45.71%). These companies were in stable operation, displaying high profitability and great growth potential. Among the 508 companies, 20 reported a two-year average net

Table 2.4 Profit and loss of Zhongguancun NEEQ listed companies from 2015 to 2019

Year	Profitable companies		Unprofitable companies		Total	
	Quantity	Proportion (%)	Quantity	Proportion (%)	Quantity	Proportion (%)
2015	574	76.03	181	23.97	755	100.00
2016	1022	71.22	413	28.78	1435	100.00
2017	978	67.63	468	32.37	1446	100.00
2018	734	59.97	490	40.03	1224	100.00
2019	626	57.91	455	42.09	1081	100.00

Source Wind, compiled by ZLCA

Table 2.5 Profit and loss of Zhongguancun NEEQ listed companies in different levels in 2019

Classification	Profitable companies		Unprofitable companies	
	Quantity	Proportion (%)	Quantity	Proportion (%)
Innovation	60	66.67	30	33.33
Base	566	57.11	425	42.89

Source Wind, compiled by ZLCA

profit of no less than 100 million yuan (see Table 2.6), and 129 of no less than 20 million yuan.

2.3.3 Size Distribution

The net profit of 44.22% Zhongguancun NEEQ Listed Companies (478) was in the range of 0–20 million yuan, which was the most concentrated one. There were 84 companies with a net profit of 20 to 40 million yuan, accounting for 7.77%; 25 of 40 to 60 million yuan, accounting for 2.31%; 14 of 60 to 80 million yuan, accounting for 1.3%; 25 of 80 million yuan and above, accounting for 2.31%. In addition, there were 455 (42.09%) companies at a loss. On the whole, most Zhongguancun NEEQ Listed Companies reported a net profit of less than 20 million yuan, and only 2.31% companies had a revenue of 80 million yuan or more. Combined with the revenue of Zhongguancun NEEQ Listed Companies, it is concluded that that Zhongguancun NEEQ Listed Companies are mainly in small and medium size.

The net profit of the NEEQ Listed Companies on base tier was mainly in the range of 0–20 million yuan and loss, with 455 and 425 companies respectively, accounting for 45.91 and 42.89%. There were 66, 142, 12 and 19 companies on base tier reporting a net profit of 20–40 million yuan, 40–60 million yuan, 60–80 million yuan, 80 million and above, accounting for 6.66%, 1.41%, 1.21% and 1.21% respectively. The net profit distribution of companies on base tier and of all Zhongguancun NEEQ Listed Companies shows the same trend.

The net profit distribution of Zhongguancun NEEQ Listed Companies on innovation tier was slightly better than that of the base tier, and the proportion of loss-making companies was far lower than that of the latter. In 2019, a total of 30 companies on innovation base reported loss, accounting for 33.33%, which is lower than the proportion of loss-making companies on base tier; There were 23 companies with a net profit ranging from 0 to 20 million yuan, accounting for 25.56%; 18 of 20 million to 40 million yuan, accounting for 20.00%; 11 of 40–60 million yuan, 2 of 60–80 million yuan and 6 of 80 million and above, accounting for 12.22%, 2.22% and 6.67% respectively (see Table 2.7). On the whole, companies with a net profit of 0–20 million yuan, 20–40 million yuan and 40–60 million yuan accounted for 57.78% of the total, and the numbers of companies in the three ranges were quite even, which shows that the profitability of companies on innovation tier was slightly better than that of those on base tier.

Table 2.6 Zhongguancun NEEQ listed companies reporting a two-year average net profit of not less than 100 million yuan in 2018 and 2019

Number	Code	Name	Tier	Industry	Net profit in 2019 (million yuan)	Net profit in 2018 (million yuan)	Average net profit (million yuan)
1	838,966.OC	MICROFUNPLUS	Base	Information transmission, Software and IT services	138.41	98.20	118.31
2	871,196.OC	Jiaoda Signal Technology	Base	Information transmission, Software and IT services	118.71	106.48	112.60
3	833,858.OC	CHINAEQUITY	Base	Financials	157.92	690.81	424.37
4	430,719.OC	TCJDCAPITAL	Base	Financials	1947.84	739.55	1343.69
5	430,005.OC	HTA	Innovation	Manufacturing	279.26	253.65	266.45
6	872,801.OC	ELEX	Base	Information transmission, Software and IT services	620.74	755.97	688.35
7	831,344.OC	3S LIFT	Innovation	Manufacturing	141.78	93.33	117.55
8	833,966.OC	TECHNOLOGY AND INVESTMENT	Base	Scientific research and technology service	105.66	101.47	103.56
9	430,277.OC	SHENGSHANG EDUCATION	Base	Leasing and Business services	173.39	97.25	135.32
10	837,747.OC	BEIJING CHANGJIANG CULTURE	Innovation	Culture, sports and entertainment industry	111.07	121.04	116.05

(continued)

Table 2.6 (continued)

Number	Code	Name	Tier	Industry	Net profit in 2019 (million yuan)	Net profit in 2018 (million yuan)	Average net profit (million yuan)
11	832,924.OC	BRIGHTSTONE	Innovation	Manufacturing	567.22	663.86	615.54
12	836,019.OC	IAT	Base	Scientific research and technology service	119.62	104.93	112.28
13	833,499.OC	CKFIL	Base	Leasing and Business services	380.40	404.07	392.24
14	430,193.OC	WIN MEDIA	Base	Information transmission, Software and IT services	96.09	107.62	101.86
15	833,819.OC	NUTRICHEM	Innovation	Manufacturing	267.12	460.88	364.00
16	834,082.OC	CNBM TECHNOLOGY CO.,LTD	Innovation	Information transmission, Software and IT services	298.05	215.84	256.95
17	835,281.OC	HANLINHUI	Base	Wholesale and Retail businesses	215.60	195.94	205.77
18	834,777.OC	I&G	Base	Leasing and Business services	830.81	1860.64	1345.73
19	870,488.OC	GUODU SECURITIES	Base	Financials	539.73	−153.54	193.10
20	834,344.OC	CHINA POST FUND	Base	Financials	152.11	164.57	158.34

Source Wind, compiled by ZLCA

Table 2.7 Distribution of net profit of Zhongguancun NEEQ listed companies in 2019

Net profit	Overall		Base		Innovation	
	Quantity	Proportion (%)	Quantity	Proportion (%)	Quantity	Proportion (%)
Loss	455	42.09	425	42.89	30	33.33
0–20 million yuan	478	44.22	455	45.91	23	25.56
20 million–40 million yuan	84	7.77	66	6.66	18	20.00
40 million yuan–60 million yuan	25	2.31	14	1.41	11	12.22
60 million–80 million yuan	14	1.3	12	1.21	2	2.22
80 million yuan and above	25	2.31	19	1.92	6	6.67

Source Wind, compiled by ZLCA

2.3.4 Summary of the Profitability of Zhongguancun NEEQ Listed Companies

The net margin of Zhongguancun NEEQ Listed Companies in continuous operation registered robust growth. On the whole, the gross profit of Zhongguancun NEEQ Listed Companies in continuous operation in 2019 has decreased slightly, and the net margin has increased over 2018. The gross margin and net margin in 2019 were 22.62% and 4.27%, respectively. In 2019, the profitability of Zhongguancun NEEQ Listed Companies increased. Specifically, the net profit of continuously listed companies on base and innovation tiers both increased, and the net margin of companies on innovation tier has increased more than that of companies on base tier. In general, the profitability of companies on innovation tier is better than that of those on base tier (see Table 2.8).

Table 2.8 Summary of the profitability of companies listed on the Zhongguancun NEEQ from 2018 to 2019

	Gross profit (billion yuan)			Net profit (billion yuan)			Gross margin (%)		Net margin (%)	
	2018	2019	Growth rate	2018	2019	Growth rate	2018	2019	2018	2019
Innovation	14.24	10.26	−27.95	2.02	1.93	−4.27	22.19	21.47	3.14	4.04
Base	39.45	36.14	−8.38	6.65	6.82	2.59	24.06	22.97	4.06	4.33
Total	53.68	46.04	13.57	8.67	8.75	0.99	23.53	22.62	3.80%	4.27

Source Wind, compiled by ZLCA

Table 2.9 ROA and ROE of Zhongguancun NEEQ listed companies from 2018 to 2019

	ROA (%)		ROE (%)	
	2019	2018	2019	2018
Innovation	2.78	2.31	4.81	4.10
Base	3.30	2.04	4.26	4.34
Total	2.66	2.09	4.3	4.29

Source Wind, compiled by ZLCA

2.4 Return on Assets and Return on Equity

In 2019, the return on assets (ROA) and return on equity (ROE[2]) of Zhongguancun's NEEQ Listed Companies were 2.66% and 4.38%, respectively. Compared with 2018, both have increased to a certain extent, indicating that companies' profitability and operation efficiency has increased.

The ROA of Zhongguancun NEEQ Listed Companies on base tier in 2019 was higher than that of the innovation tier, and the ROE was lower than that of the latter (see Table 2.9). This can also reflect that the net assets of companies on innovation tier account for a relatively high proportion of total assets, which means more secure financial structure and lower risks. At the same time, compared with 2018, the ROA and ROE of Zhongguancun NEEQ Listed Companies in 2019 have both increased, indicating that the profitability of the innovation-level companies has been growing.

Most Zhongguancun's NEEQ listed Companies reported an ROA of 0–20%. There were 558 of them, accounting for 51.62%. Followed were companies with an ROA of −20% to 0, namely 277 companies, accounting for 25.65%. In addition, there were 59, 9 and 178 companies with a ROA of 20–40%, more than 40%, and lower than −20%, accounting for 5.46%, 0.83%, and 16.47% respectively. Overall, there were 455 companies with a negative ROA, accounting for 42.09% (increased compared over 2018). The overall operating conditions and profitability of Zhongguancun NEEQ Listed Companies needs to be improved (see Fig. 2.5).

With 42 companies holding negative net assets in 2018 and 2019 excluded, among the remaining 1039 Zhongguancun NEEQ Companies, most of them reported an ROE of 0–20%, accounting for 44.56% (463 companies); followed were those with an ROE of −20% to 0, namely 182 companies, accounting for 17.52%. There were 124, 111, 113 and 46 with a return on total assets of less than −40%, −40% to −20%, 20–40%, and greater than or equal to 40%, accounting for 11.93%, 10.68%, 10.88%, 4.43%, respectively (see Fig. 2.6). On the whole, there were 622 Zhongguancun NEEQ Listed Companies with an ROE greater than 0, accounting for 57.54%. These companies show high profitability and competitiveness.

[2]ROE = Net profit/average net assts × 100%; ROA = Net profit/average total assets × 100%.

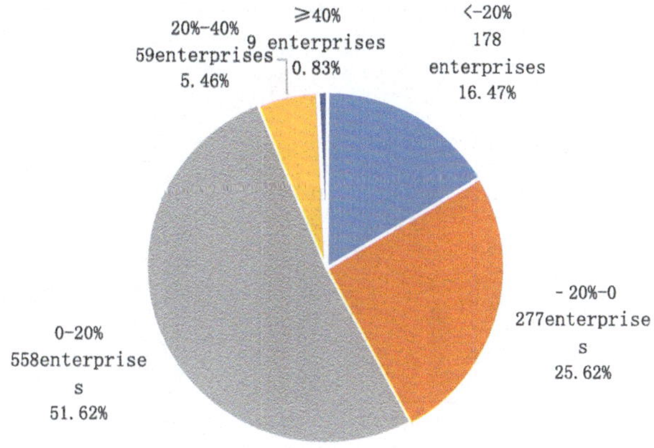

Fig. 2.5 Distribution of ROA of Zhongguancun NEEQ listed companies in 2019. *Source* Wind, compiled by ZLCA

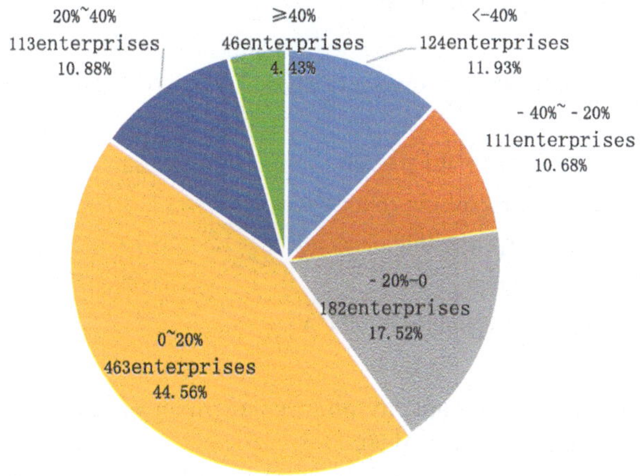

Fig. 2.6 Distribution of ROE of Zhongguancun NEEQ companies in 2019. *Source* Wind, compiled by ZLCA

2.5 Period Costs

The period expenses including sales expenses, management expenses and financial expenses of the 1081 Zhongguancun NEEQ Listed Companies totalled 34 billion yuan in 2019, a year-on-year decrease of 16.41%. The average period costs in 2019 was 31.45 million yuan, a year-on-year decrease of 5.6%. Among them, sales expenses were 12.824 billion yuan, a year-on-year decrease of 296 million yuan; management expenses were 17.053 billion yuan, a year-on-year decrease of 0.51%; financial expenses were 3.785 billion yuan, a year-on-year increase of 7.64%. The period expenses including sales expenses, management expenses and financial expenses of the 1035 Zhongguancun NEEQ Listed Companies that continued to operate in 2018 and 2019 reached 33.662 billion yuan in 2019, a year-on-year decrease of 0.62%. The revenue of companies in continuous operation rose sharply, while their sales and management expenses dropped relatively. It shows that

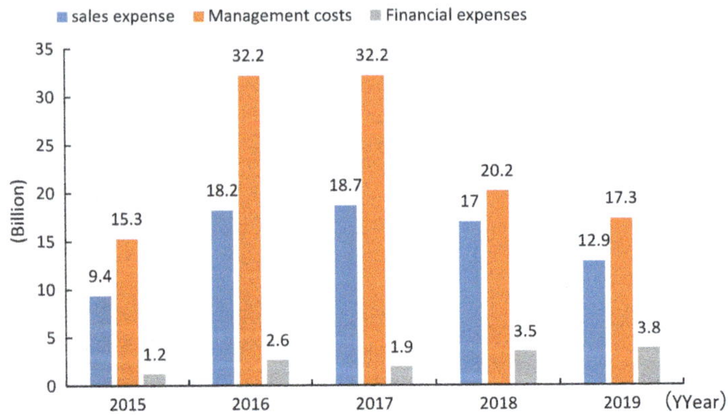

Fig. 2.7 Period expenses of Zhongguancun NEEQ listed companies from 2015 to 2019. *Source* Wind, compiled by ZLCA

Table 2.10 Period expenses of Zhongguancun NEEQ listed companies in continuous operation

	2019	2018	Growth rate (%)
Sales expenses (billion yuan)	12.82	13.22	−2.96
Management expenses (billion yuan)	17.05	17.14	−0.51
Financial expenses (billion yuan)	3.79	3.52	7.64
Total	33.66	33.87	−0.62

Source Wind, compiled by ZLCA

the marketing capacity and management efficiency of these companies have been effectively improved (see Fig. 2.7).

Similar to the overall tendency of Zhongguancun NEEQ Listed Companies, companies in continuous operation reported a certain decline in the sales and management expenses in 2019. However, the financial expenses of companies in continuous operation and the whole Zhongguancun NEEQ Listed Companies both registered a 8% increase. Among them, the financial expenses and interest expenses of companies in continuous operation in 2019 totalled 3.729 billion yuan, accounting for 98.52% of the financial expenses of the Zhongguancun NEEQ Listed Companies in 2019, a year-on-year increase of 2.61% (see Table 2.10).

In 2019, the proportion of management expenses (50.88%) was the higest, followed by that of sales expenses, namely 37.94%. Financial expenses were in a relatively small proportion of 11.18%, but increased by 2.5% compared to 2018 (8.68%) (see Fig. 2.8).

In 2019, the management expense ratio[3] and sales expense ratio[4] of Zhongguancun NEEQ Listed Companies decreased over the previous year, and have been overall

[3]Management expense ratio = management expenses/revenue × 100%.

[4]Sales expense ratio = sales expenses/revenue × 100%.

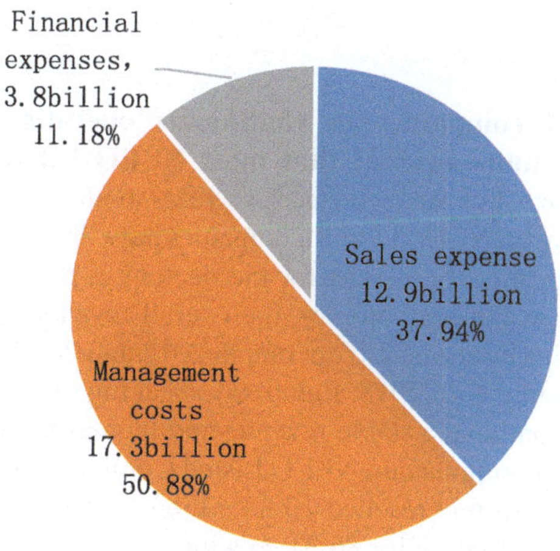

Fig. 2.8 Proportion of period expenses of Zhongguancun NEEQ listed companies in 2019. *Source* Wind, compiled by ZLCA

declining over the past five years, which indicates that the overall management efficiency and marketing efficiency of the companies have been continuously improving. In 2019, the financial expense ratio of Zhongguancun NEEQ Listed Companies increased from 1.48 in 2018 to 1.84 in 2019, and has continued to rise for five consecutive years (see Fig. 2.9).

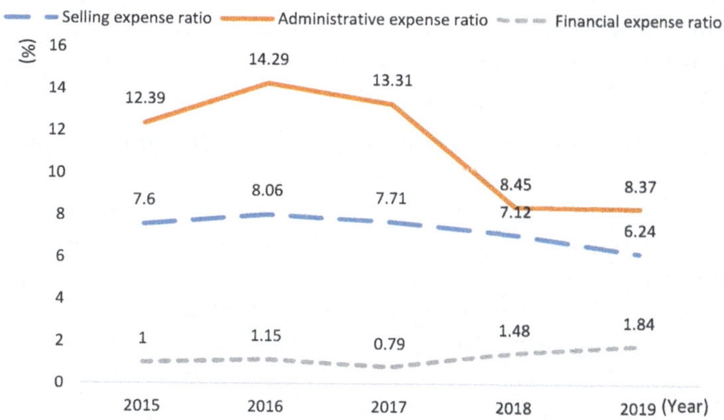

Fig. 2.9 Changes in the period expense ratio of Zhongguancun NEEQ listed companies from 2015 to 2019. *Source* Wind, compiled by ZLCA

2.6 Summary

The profitability of companies in continuous operation has significantly increased, and it is more evident that most NEEQ listed companies are in small and medium size. For three consecutive years, both total revenue and average revenue of Zhongguancun NEEQ Listed Companies have dropped, so have the overall gross profit and net profit. The reason for the decline in profitability is mainly the delisting of companies. On the contrary, the overall revenue, gross profit, and net profit of companies that continued to operate in 2018 and 2019 have all increased to a certain extent compared with 2018. Relatively speaking, NEEQ Listed Companies in continuous operation have a stable revenue and increasing profitability. In addition, there were 532 Zhongguancun NEEQ Listed Companies with a revenue of less than 50 million (total revenue reached 10.38 billion yuan, accounting for 49.21%), and 933 companies (accounting for 86.31%) with a net profit of less than 20 million yuan. This indicates that a large number of NEEQ Listed Companies are in small and medium size, while large-scale companies in a small number generated most of the revenue. Specifically, the 27 companies taking up 2.5% of the total had a revenue of 95.559 billion yuan overall, accounting for 46.23% of the total revenue.

The marketing capacity and management efficiency of Zhongguancun NEEQ Listed Companies have been improved year by year. The sales expenses and management expenses of Zhongguancun NEEQ Listed Companies have dropped. The period expenses of the 1,035 Zhongguancun NEEQ Listed Companies that continued to operate in 2018 and 2019 reached 33.662 billion yuan, a year-on-year decrease of 0.62%. The revenue of companies in continuous operation has risen sharply, while their sales and management expenses have decreased, suggesting that the marketing capacity and management efficiency of companies in continuous operations have been effectively improved.

The overall profitability has increased, and the profitability of companies on innovation tier has grown. In 2019, both the ROA and ROE of Zhongguancun NEEQ Listed Companies have increased to a certain extent, reflecting the increase in profitability and management efficiency. The ROA and ROE of companies on innovation tier have both increased to a certain extent, and the average gross profit and average net profit of these companies are better than those on base tier. The profitability of Zhongguancun NEEQ Listed Companies has continued to grow.

Chapter 3
Research Report on Innovation Capability of Zhongguancun NEEQ Listed Companies in 2019

Abstract This report studies and analyzes the innovation capabilities of Zhongguancun NEEQ enterprises with innovation input and R&D output as the main evaluation dimensions. The research results show that both total R&D expenses and average R&D expenses have declined in 2019, mainly due to the decrease in the number of Zhongguancun NEEQ enterprises. The intensity of R&D is slightly lower than last year, mainly due to the impact of the Sino-US trade war and the downturn in macro fundamentals, which makes the R&D strategy of enterprises relatively conservative. There is a relatively significant positive correlation between corporate profitability and R&D investment. The transformation ability of enterprises' innovation achievements has been effectively improved, and the number of patent applications and patent grants in enterprise innovation achievements has been steadily improving; the number of PCT patent applications has risen sharply to 25, an increase of 47% compared with 2018. The results of innovation output indicate that the international strategic layout of Zhongguancun NEEQ Enterprises has been carried out efficiently, and theirs innovation ability has been recognized by the international market.

Keyword Zhongguancun NEEQ · R&d input · R&d output

3.1 Innovation Input

3.1.1 R&D Expenses of Zhongguancun NEEQ Listed Companies in 2019

Under the dual pressure of Sino-US economic and trade war and economic downturn, the scientific innovation is vital to the survival and developments of the companies. For enterprises, the increase in R&D efforts, improving R&D quality and enhancing the core competitiveness of enterprises are the one of the most important strategies related to the long-term development. In 2019, a total of 862 Zhongguancun NEEQ listed companies have disclosed the R&D expenses, accounting for 79.74%. The disclosed total R&D expenses is 8.45 billion, there is a 15.75% YOY decrease.

Zhongguancun Listed Companies Association, *The Growth Report of Zhongguancun NEEQ Listed Companies (2020)*, Current Chinese Economic Report Series,
https://doi.org/10.1007/978-981-33-6819-4_3

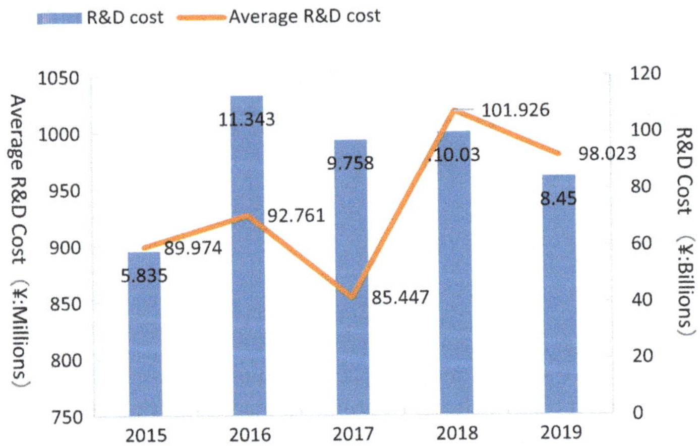

Fig. 3.1 Zhongguancun NEEQ listed companies' R&D cost from 2014 to 2018. *Source Wind*, complied by ZLCA

There are 806 208–2019 continuous operating companies have disclosed the R&D expenses that totals 8.244 billion yuan. This amount is roughly the same to the 2018, which shows the decrease in R&D expenditure of Zhongguancun NEEQ listed companies is due to the reduction in the number of the companies. (See Fig. 3.1). The growth of R&D expenses of these 806 enterprises is analyzed in detail, there are 452 companies realized a positive growth, accounting for 56.08%; The 354 companies realized a negative growth, accounting for 43.92%.

2019 is the outbreak year of Zhongguancun NEEQ enterprises for board transfer, but the transfer of high-quality NEEQ enterprises has not affected the change of R&D expenditure scale of head enterprises. In 2019, the 50 million yuan above R&D expenditure of the NEEQ companies disclosed do not have a significant change, while the number of enterprises with R&D expenses between 1 and 10 million yuan decreased more. Specifically, the R&D expenses above 100 million-*yuan*, accounting for 0.70% of the total number of enterprises that disclose R&D expenses. There are 14 enterprises with R&D expenses ranging from 50 to 100 million-*yuan*, accounting for 1.62%. There are 227 enterprises with R&D expenses ranging from 10 to 50 million-*yuan*, accounting for 26.33%. The number of enterprises with R&D expenses between 5 and 10 million yuan decreased to 178, accounting for 20.65%. There are 333 enterprises with R&D ranging from 1 to 5 million-*yuan*, accounting for 38.63%. The number of enterprises with R&D expenses below 1 million yuan increased to 104, accounting for 12.07%. The statistical results show that the reduction of the total R&D expenses of Zhongguancun NEEQ enterprises in 2019 is mainly reflected in the enterprises with small R&D investment base, but has little impact on the enterprises with high investment (see Fig. 3.2).

R&D intensity refers to the proportion of R&D investment in business income, which is an importantindext to measure innovation investment of enterprises. International experience shows that enterprises with high R&D intensity have higher profit

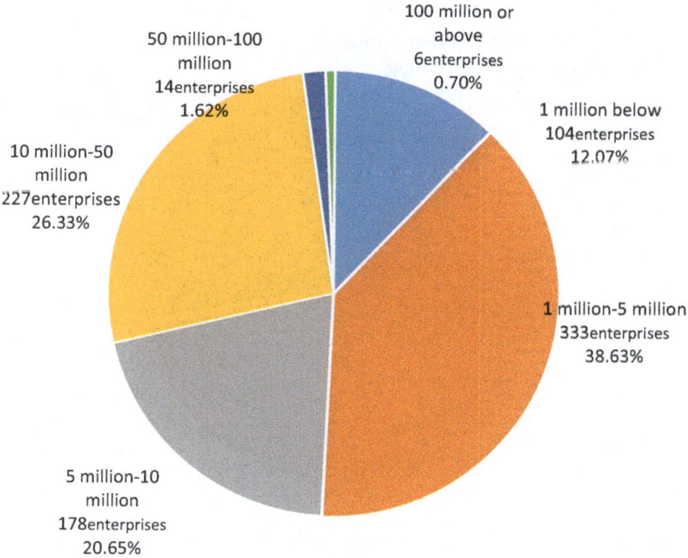

Fig. 3.2 Zhongguancun NEEQ listed companies R&D cost distribution. *Source Wind*, complied by ZLCA

margins. In 2019, the average R&D intensity of Zhongguancun NEEQ enterprises is 4.7%, which was slightly lower than that in 2018,[1] but still higher than the average R&D intensity of national NEEQ enterprises[2] and the whole society in 2019.[3]

International experience shows that more than 10% of R&D intensity is at a high level of R&D investment worldwide, and such enterprises are generally considered to have sufficient R&D competitiveness advantages. According to the standard, the R&D intensity of Zhongguancun NEEQ enterprises is considerable. There are 368 companies are high R&D intensity in 2019, there were 368 enterprises with high R&D intensity, accounting for 42.7%, up 2.7% YOY. It shows that enterprises with high intensity R&D are less affected when macro fundamentals go down. Zhongguancun NEEQ leading companies have increased R&D investment by contrarian and improved its core competitiveness (see Fig. 3.3).

In 2018, the total R&D investment of the top 30 companies in Zhongguancun NEEQ reached 2.481 billion-*yuan*, accounting for 29.36% of the total R&D investment of Zhongguancun NEEQ companies.

[1]In 2018, the Zhongguancun listed companies average R&D intensity was 5.55%.

[2]In 2019, the national NEEQ enterprises average R&D intensity was 4.15%.

[3]In 2019, the ratio of R&D expenditure to GDP was 2.19%.

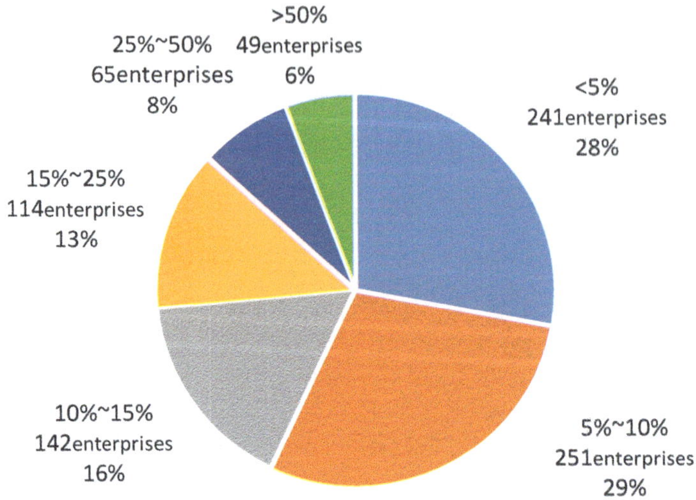

Fig. 3.3 Zhongguancun NEEQ listed companies R&D intensity distribution. *Source Wind*, complied by ZLCA

3.1.2 Relationship Between R&D Investment and Profitability

The depth of R&D of the companies differentiate the products, and highly innovative products often bring a deep moat to the competitiveness to companies. Data show that there is a positive correlation between R&D intensity and gross profit margin of Zhongguancun NEEQ companies, and companies with high R&D intensity have higher gross profit margin. Specifically, the average gross profit margin of more than 50% companies reached 61.84% in 2019, which is 3.86-time s of the average gross profit margin of 16.03% of enterprises with R&D intensity below 5%. The R&D brings uncertainties, companies must be careful when making R&D decisions. The companies with too low R&D investment cannot gain long- term competitiveness, while too high R&D investment brings certain financial pressure to business operations. Reasonable R&D intensity that meets the conditions of companies' own scale and strength should be pursued by companies (see Fig. 3.4).

3.2 Innovation Output

3.2.1 Patent Application and Ranking

Patent application is one of the important indicators to measure the innovation output of companies, which refers to the number of patent applications for technical inventions accepted by domestic and foreign patent institutions., including the number

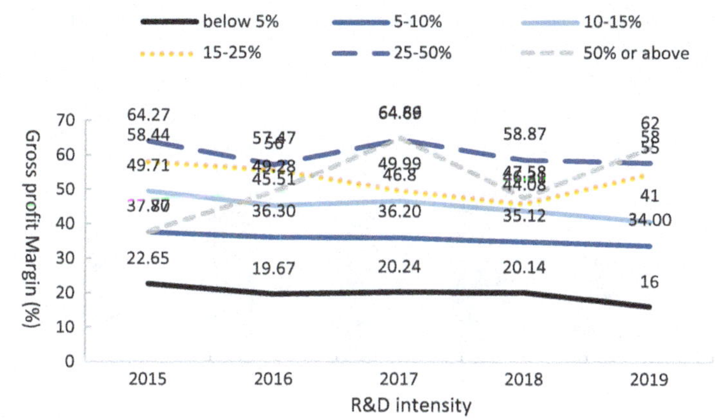

Fig. 3.4 Gross profit margin and R&D cost from 2015 to 2019. *Source Wind*, complied by ZLCA

of invention patent applications, utility model patent applications and design patent applications. The more patent applications an company has, the more it can show that the enterprises has an active innovation atmosphere and strong innovation ability.

From the perspective of the total number of patent applications, the number of patent applications for NEEQ was 2362 in 2019, which was 29.81% lower than that in 2018. By analyzing the enterprises that have applied for patents for two consecutive years, it is found that 332 Zhongguancun companies that applied for patents for two consecutive years from 2018 to 2019 have applied for 2316 patents, up 9.19% YOY. This set of data shows that the dual pressures of Sino-US economic and trae war and economic downturn have not affected the patent research and development process of the NEEQ companies. From the perspective of the average number of patent applications, a total of 339 companies applied for patents in 2019, with an average of 6.97 patents per company, down 17.51% from 8.45 in 2018. However, the number of enterprises that reached the average number of patent applications increased from 84 in 2018 to 103 in 2019, a YOY increase of 22.62%. From the perspective of patent category, the number of invention patent application in 2019 was 914, accounting for 38.70% of the total number of patent applications. The number of utility model patent applications was 1158, accounting for 49.03%. The number of design patent appli cations was 290, accounting for 12.27% (Fig. 3.5). The invention patent application as the one with the highest value and the strictest examination link, the application amount is less than 40%, indicating that the R&D quality of Zhongguancun NEEQ companies needs to be improved.

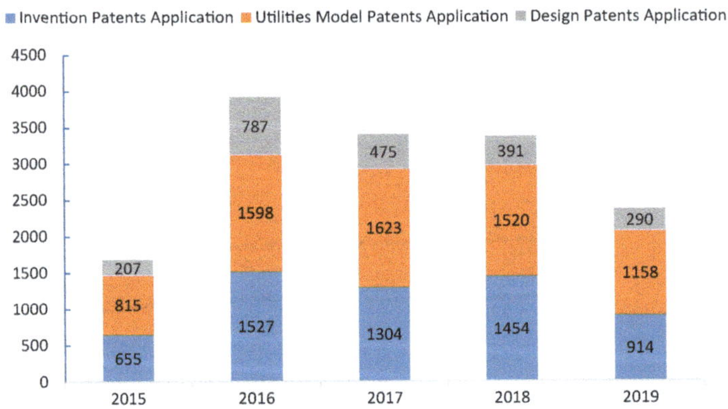

Fig. 3.5 Zhongguancun NEEQ listed companies patents application status from 2015 to 2019. *Source* Zhongguancun Intellectual Property Promotion Bureau

3.2.2 Patents Authorization and Ranking

The principle of granting a patent right is that the patent has novelty, creativity and practicability. The patent is protected by law and is the private property of the enterprise after being authorized. The higher the patent authorization, the stronger the comprehensive strength of companies' innovation output, as it is the one of the main indicators to measure the ability, level and quality of companies' innovation output. The total number of patents granted by Zhongguancun companies was 1450 in 2019, down from 2144 in 2018. There are 305 invention patents, 918 utility model patents and 227 design patents were authorized in 2019, accounting for 21.03%, 63.31% and 15.66% respectively.

In comparison with the percentage of different types of patent applications, it is found that the proportion of invention patents granted by companies listed on Zhongguancun NEEQ is relatively low, which may be related to the long examination period and low examination pass rate of invention patents. In view of this, the Zhongguancun NEEQ listed companies are suggested that they can cooperate with the external resources, such as Zhongguancun universities, scientific research institutions and technical service institutions to further enhance their innovation capabilities. At the same time, the relevant departments are expected to speed up patent examination and approval, and help enterprises use patents to maintain their core competitiveness (See Fig. 3.6).

3.2.3 Number and Ranking of Valid Patents

A valid patent is a patent that is still valid after the patent application is authorized. The longer an invention patent is maintained, the longer it will create economic

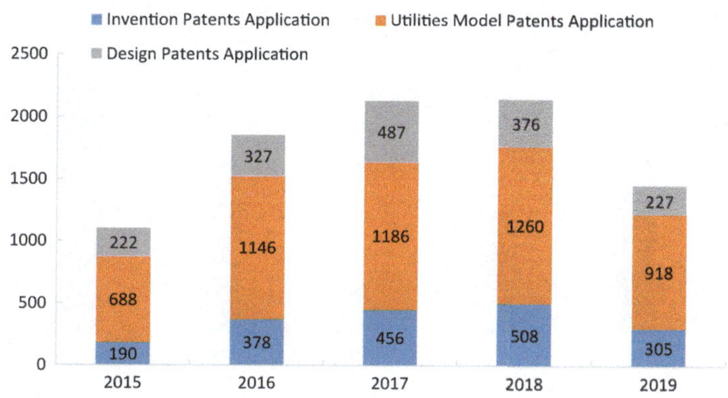

Fig. 3.6 Number of patents of Zhongguancun NEEQ listed companies from 2015 to 2019. *Source* Zhongguancun Intellectual Property Promotion Bureau

benefits. Data shows that only 40% of domestic invention patents have been maintained for more than 10 years, while the 82.2% of foreign invention patents have been maintained for more than 10 years. That is the average maintenance time of domestic invention patents is lower than that of international invention patents, reflecting that the innovation output of domestic enterprises is mainly short and fast. Until 31st December 2019, a total of 373 companies in Zhongguancun NEEQ had had valid invention patents, totaling 2452 which has a slight increase compared with 2423 in 2018 (see Fig. 3.7).

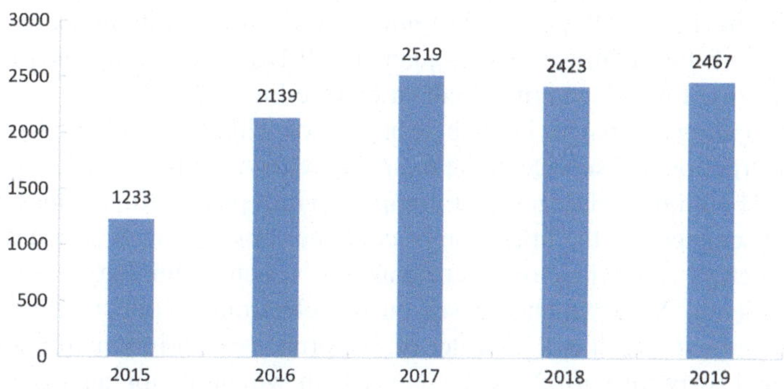

Fig. 3.7 Zhongguancun NEEQ listed companies with valid invention patents from 2015 to 2019. *Source* Zhongguancun Intellectual Property Promotion Bureau

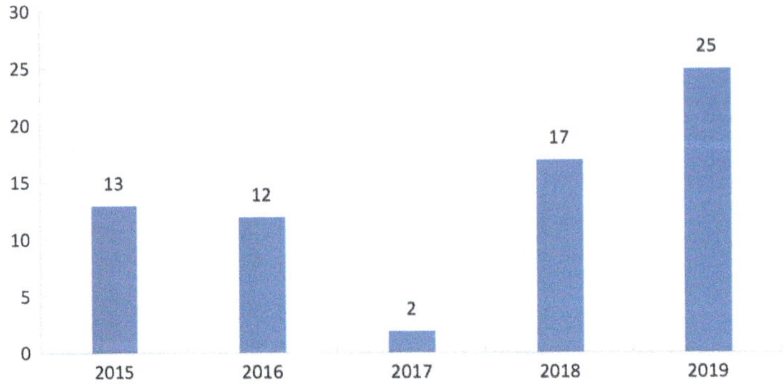

Fig. 3.8 Number of PCT application of Zhongguancun NEEQ companies from 2015 to 2019. *Source* Zhongguancun Intellectual Property Promotion Bureau

3.2.4 PCT Patent Application

PCT, abbreviated for Patent Cooperation Treaty, is an international cooperation treaty in the patent field, which aims to simplify the method of applying for invention patent protection in many countries and make the application process of patent protection more economic and effective. The increase in the number of PCT applications by companies indicates that the company's invention patents are gradually globalized and widely recognized in the international market.

In 2019, the number of PCT applications of Zhongguancun NEEQ companies reached a new high, reaching 25 pieces, an increase of 47% compared with 17 in 2018. It reflects the growth potential of innovation ability of Zhongguancun NEEQ companies, and the quality of Zhongguancun NEEQ listed companies is moving from localization towards internationalization (see Fig. 3.8).

The outstanding companies include Strong Union which are dedicated to the information construction and service in the financial field with 5 pieces, Elex which is an international boutique mobile network company with 4 pieces, Rui Li Heng Yi Logistics which is engaged in the operation of container brackets with 3 pieces, Hanvon Pentech which is a provider of electromagnetic touch technology with 2 pieces, Aerospace Chang Xing which focuses on fire electronic products with 2 pieces, General Treatment which is a provider of comprehensive solutions for solid waste treatment and disposal with 2 pieces, Nutrichem which a supplier of agricultural chemical products with 2 pieces, Comfort Technologies which provides technical services for central air conditioning terminal equipment and building automation system, Longtel which provides professional integrated telecommunication services for customer with 2 pieces, Yimo which provides integrated solutions for simulated teaching products and modern medical simulation centers with 1 piece and

Table 3.1 Number of PCT application of Zhongguancun NEEQ listed companies in 2019

Bond number	Bond name	Industry	Number of PCT application
430,152.OC	Strong Union	Information transmission, Software and IT services	5
872,801.OC	Elex	Information transmission, Software and IT services	4
831,275.OC	Rui Li Heng Yi Logistics	Manufacturing	3
833,819.OC	Nutrichem Company	Manufacturing	2
873,062.OC	Hanvon Pentech	Manufacturing	2
835,565.OC	General Treatment	Management of Water Conservancy, Environment and Public Facilities	2
430,209.OC	Comfort Technologies	Manufacturing	2
835,697.OC	Aerospace Chang Xing	Manufacturing	2
831,091.OC	Jingyeyuan	Manufacturing	1
832,649.OC	Yimo	Manufacturing	1
839,433.OC	Longtel	Information transmission, Software and IT services	1

Source Zhongguancun Intellectual Property Promotion Bureau

Jingyeyuan which is a provider of building materials and refractory materials (Table 3.1). PCT's patent applications are mainly concentrated in manufacturing information and information transmission, software and information transmission, software and information technology services, water conservancy, environment and public facilities management.

Chapter 4
Analysis of the Solvency and Operating Capability of Zhongguancun NEEQ Listed Companies in 2019

Abstract This paper measures the overall solvency of Zhongguancun NEEQ enterprises through indicators such as asset-liability ratio, cash flow debt ratio, current ratio, etc; and presents the operating capacity of Zhongguancun NEEQ enterprises through indicators such as account receivable turnover days, inventory turnover days, and total asset turnover days. Through corresponding research,we found that the overall debt ratio of Zhongguancun NEEQ enterprises is relatively low and the solvency is good; the fixed assets turnover days, total assets turnover days and other indicators are excellent, and the operating capacity is strong.

Keyword NEEQ enterprises · Solvency · Operating ability

4.1 Analysis of the Solvency of Zhongguancun NEEQ Listed Companies in 2019

4.1.1 Overview of Solvency

Solvency reflects a company's ability to repay debt. Although it does not have a direct connection with the company's profit growth, it is important for the company's survival and development, and is a basic element to measure the company's growth. For the vast majority of NEEQ companies that are in the start-up and growing stages, they need sufficient funds to support their rapid growth, which is accompanied by higher financial risks. If the NEEQ companies are not fully aware of the importance of solvency, it is easy for them to run into improper use of funds, which will cause the company to be unable to repay the due debts, thus falling into financial distress. Or they will have to deal with the increase in funding costs and financial expense ratios, which reduces their profitability. Therefore, this chapter intends to analyse the overall debt solvency of Zhongguancun NEEQ Listed Companies to reflect the financial status of them so as to provide suggestions for companies to adjust their capital structures and rationally plan the proportion of debt financing and offer policy basis for relevant government departments to prevent systemic risks in business operations.

Zhongguancun Listed Companies Association, *The Growth Report of Zhongguancun NEEQ Listed Companies (2020)*, Current Chinese Economic Report Series,
https://doi.org/10.1007/978-981-33-6819-4_4

In a financial analysis system, solvency is often divided into short-term solvency and long-term solvency. Short-term solvency focuses on the ability of enterprises to repay current liabilities, while long-term solvency focuses capital structure and the ability to repay long-term debt. For short-term solvency, this report uses more important indicators such as current ratio, quick ratio, and cash ratio, for long-term solvency, it uses asset-liability ratio, cash flow and debt ratio. Through the calculation and analysis of these indicators, it is hoped to overall present the solvency of Zhongguancun NEEQ Listed Companies, so as to provide suggestions for the sustainable and healthy growth of them.

4.1.2 Analysis of Long-Term Solvency

1. Capital Structure

   (1) Total assets
   On December 31st, 2019, the total assets of Zhongguancun NEEQ Companies were worth 368.225 billion yuan, which was a decrease of 16.86% from 2018 (442.88 billion yuan) due to the reduction in the number of companies. Of the total assets, the assets of companies on innovation tier were 71.700 billion yuan, accounting for 19.47%. The average assets of Zhongguancun NEEQ Companies were 341 million yuan, and there were 149 enterprises whose assets were higher than the average; on innovation tier, the average assets were 797 million yuan, and there were only 15 companies whose assets were higher than the average. The assets distribution of Zhongguancun NEEQ Companies shows that there is a clear polarization in the size of enterprise assets, and there are various sizes of Zhongguancun NEEQ Companies.

   (2) Total liabilities
   As of December 31st, 2019, the total liabilities of Zhongguancun NEEQ Companies were worth 179.574 billion yuan. The sector of the companies on innovation tier were 30.544 billion yuan, accounting for 17.01%. The average liabilities of Zhongguancun NEEQ Companies were 166 million yuan, a decrease of 10.75% over 2018 (186 million yuan); the average debt of companies on innovation tier was 339 million yuan, an increase of 6.27% from 2018 (319 million yuan). The change in average liabilities initially is an early sign of a decrease in the overall debt of Zhongguancun NEEQ Listed Companies but a slight increase in the debt of quality companies.

   (3) Asset-liability ratio
   At the end of 2019, the overall asset-liability ratio of Zhongguancun NEEQ Companies was 48.77%, and that of companies on innovation tier was 42.60%. For all NEEQ Companies in China, the overall asset-liability ratio was 58.00%. The overall asset-liability ratio of Zhongguancun NEEQ

Companies was within the reasonable range of 40–60%, and was about 10 percentage points lower than that of all NEEQ companies, which indicates lower financial risks. Compared with 2018, the overall asset-liability ratio of Zhongguancun NEEQ Companies dropped by 2.60 percentage points, and that of companies on innovation tier by 1.29 percentage points, but the overall asset-liability ratio of NEEQ companies across the country increased by 1.82 percentage points. This shows that the overall financial risks of the NEEQ companies in Zhongguancun have been reduced, and are much lower than those of all NEEQ companies nationwide.

Of all Zhongguancun NEEQ Listed Companies, there were 592 companies with an asset-liability ratio below 40%, accounting for 54.76%, 253 with an asset-liability ratio between 40 and 60%, accounting for 23.40%, 197 with an asset-liability ratio between 60 and 100%, accounting for 18.22%, 39 with an asset-liability ratio above 100%, accounting for 3.61%. It is worth noting that an asset-liability ratio higher than 100% means that the company is facing serious financial difficulties. The proportion of companies of this kind among Zhongguancun NEEQ Companies increased compared with 2018 (1.96%), which means that more companies have financial risk problems. It is generally believed that the reasonable range of a company's asset-liability ratio is 40% to 60%. The asset-liability ratio of about a quarter of Zhongguancun NEEQ Companies was within the reasonable range, and that of more than half was less than 40%.Therefore, the overall debt ratio of Zhongguancun NEEQ Listed Companies was not high, the financial risks were low, which means their long-term solvency is secured; but at the same time, there are early signs showing that half of the enterprises have insufficient debt financing and their operating efficiency is limited (see Fig. 4.1).

2. Cash Flow to Debt Ratio[1]

In order to further measure the long-term solvency of Zhongguancun NEEQ Companies, this chapter studies cash flow to debt ratio to reflect the ability of Zhongguancun NEEQ Companies to repay all debts using the net cash generated from business operations. The higher the ratio is, the stronger the ability to repay the total debt.

Data show that in 2019 the overall cash flow to debt ratio of Zhongguancun NEEQ Companies was 4.70%, and that of companies on innovation tier was 6.05%. The number of all NEEQ companies nationwide during the same period was 7.65%. This shows that the ability of Zhongguancun NEEQ Companies to use net cash flow generated from operations to repay all debts is slightly lower than all NEEQ companies nationwide.

In terms of distribution, in 2019, 462 Zhongguancun NEEQ Companies (accounting for 42.74%) had a negative net cash flow from operations, resulting in a cash flow-to-debt ratio that was also less than zero. This means that these companies did not have the ability to use the net cash obtained from operating

[1]Cash flow to debt ratio = (Generated net cash flow from operations/total debt) × 100%.

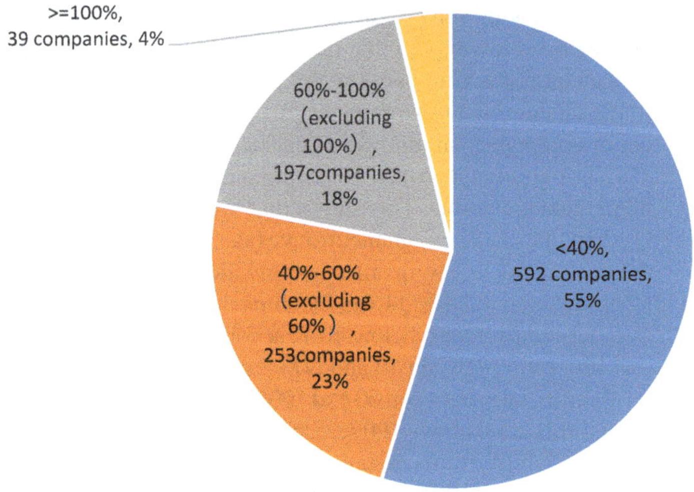

Fig. 4.1 Distribution of asset-liability ratio of Zhongguancun NEEQ companies in 2019. *Source* Wind, compiled by ZLCA

activities to repay debts in 2019. There were 271 companies (accounting for 25.07%) with a cash flow to debt ratio of 0–20% (not including 20%), which means a quite weak ability to repay total debt; 176 (accounting for 16.28%) 20–50% (excluding 50%); 99 (accounting for 9.16%) 50–100% (excluding 100%), which means that these companies' net cash flow generated from operations can repay at least one-half of the total debt; another 73 (accounting for 6.75%) more than 100%, showing strong ability to repay debt.

Compared with NEEQ companies nationwide, companies in Zhongguancun NEEQ that could not repay debt with the cash generated from operations were in the majority and the proportion of those with a cash flow to debt ratio of 0–50% (excluding 50%) is lower, and the proportion of those with a cash flow to debt ratio of more than 50% is slightly higher. The above comparisons show that compared with the polarization among Zhongguancun NEEQ Companies is more obvious than all NEEQ enterprises, and the proportion of companies with weak or general debt solvency in all Zhongguancun NEEQ companies is smaller (see Table 4.1).

4.1.3 Analysis of Short-Term Solvency

1. Short-term Debt Stock Ratio

   (1) Current ratio

   The current ratio is the ratio of current assets to current liabilities, indicating how much current assets are used for debt servicing for every 1 yuan of

Table 4.1 Comparison of cash flow to debt ratio between NEEQ Companies in Zhongguancun and Nationwide in 2019

	Zhongguancun		Nationwide	
	Quantity	Proportion (%)	Quantity	Proportion (%)
<0	462	42.74	2612	31.68
0–20% (excluding 20%)	271	25.07	2805	34.02
20–50% (excluding 50%)	176	16.28	1589	19.27
50–100% (excluding 100%)	99	9.16	732	8.88
>= 100%	73	6.75	508	6.16

Source Wind, compiled by ZLCA

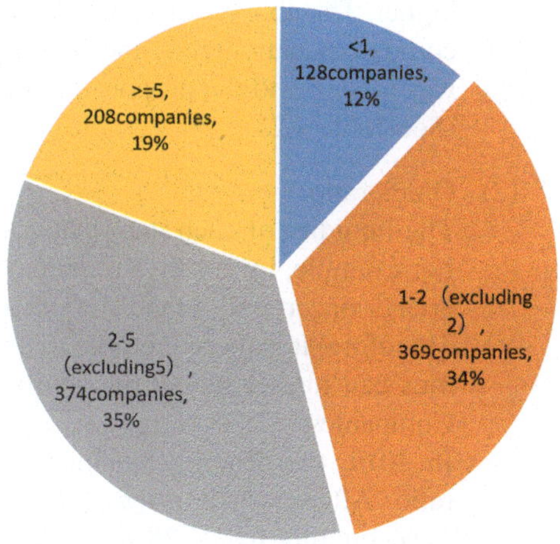

Fig. 4.2 Distribution of current ratio of Zhongguancun NEEQ companies in 2019. *Source* Wind, compiled by ZLCA

current liabilities. The current ratio is a relative number, and a value that is too low indicates that the short-term solvency of the company is insufficient, and if the value is too high, and the company may have the problem of idle funds. Generally speaking, the minimum current ratio that is considered reasonable for production enterprises is 2.

In 2019, among Zhongguancun NEEQ Companies, the current ratio of two companies was missed. With the two excluded, 128 companies reported a current ratio below 1, accounting for 11.86%; 369 companies between 1 and 2, accounting for 34.20%; 374 companies between 2 and 5, accounting for 34.66%; 208 companies above 5, accounting for 19.28%.The above data show that more than half of Zhongguancun NEEQ Companies have a current ratio of 2 or higher, which means high liquidity (see Fig. 4.2).

The proportion of NEEQ companies[2] with a current ratio of less than 1 in Zhongguancun was much lower than that across the nation, the proportion of

[2]Among all NEEQ companies in China, 18 companies did not report quick ratio.

Table 4.2 Comparison of the current ratio of NEEQ companies in Zhongguancun and nationwide in 2019

	Zhongguancun		Nationwide	
	Quantity	Proportion (%)	Quantity	Proportion (%)
<1	128	11.86	1317	16.01
1–2 (excluding 2)	369	34.20	3369	40.95
2–5 (excluding 5)	374	34.66	2551	31.00
>= 5	208	19.28	991	12.04

Source Wind, compiled by ZLCA

companies with a current ratio higher than 2 in Zhongguancun is also much higher. The above comparisons show that compared with the whole NEEQ companies nationwide, the liquidity of Zhongguancun NEEQ companies was higher, and the proportion of companies with more liquidity is greater (See Table 4.2).

(2) Quick ratio

The liquidity of each item that constitutes current assets is very different. For example, inventory and prepayments are non-quick assets, and there is greater uncertainty in the amount and time of turning such assets into cash. Therefore, it is more credible to use quick assets to repay debt. Quick ratio thus can further reflect the short-term solvency of Zhongguancun NEEQ Companies.

In 2019, two Zhongguancun NEEQ Companies reported missing quick ratio. After excluding them, 262 companies reported a quick ratio below 1, accounting for 24.28%; 335 companies between 1 and 2 (excluding 2), accounting for 31.05%; 258 companies between 2 and 4 (excluding 4), accounting for 23.91%; 224 companies above 4, accounting for 20.76%. It is generally believed that a quick ratio of 1 is the best. From the above data, it can be seen that nearly 80% of the enterprises in Zhongguancun NEEQ Companies had a quick ratio higher than 1, indicating that they have a strong ability to use quick assets to repay current liabilities (see Fig. 4.3).

Compared with the distribution of the quick ratio of the NEEQ companies across the country, it can be found that the proportion of NEEQ companies with a quick ratio of less than 1 in Zhongguancun was higher than that of all NEEQ companies. The proportion of companies with a quick ratio of 1–4 in Zhongguancun was lower than that across the nation; but for companies with a quick ratio higher than 4, the proportion in Zhongguancun was higher. This shows that there were fewer companies in Zhongguancun with a reasonable current ratio. Therefore, the allocation structure of quick assets of Zhongguancun NEEQ companies needs to be improved (see Table 4.3).

At the same time, the liquidity of accounts receivable has an important impact on the credibility of the quick ratio. Bank loans tightened in 2019,

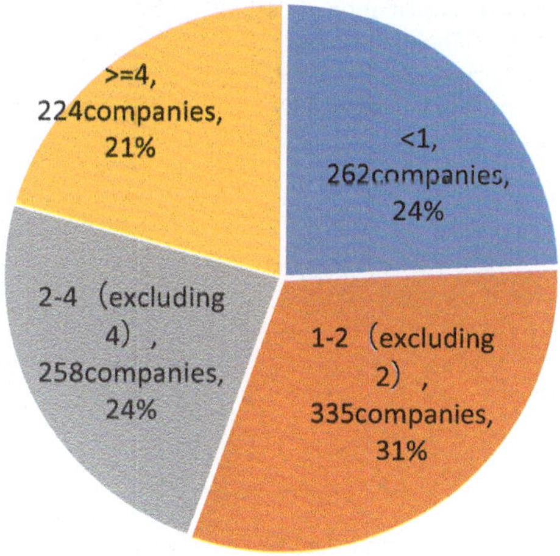

Fig. 4.3 Distribution of quick ratios of Zhongguancun NEEQ companies in 2019. *Source* Wind, compiled by ZLCA

Table 4.3 Comparison of quick ratio of NEEQ companies in Zhongguancun and nationwide in 2019

	Zhongguancun		Nationwide	
	Quantity	Proportion (%)	Quantity	Proportion (%)
<1	262	24.28	1317	16.01
1–2 (excluding 2)	335	31.05	3369	40.95
2–4 (excluding 4)	258	23.91	2200	26.74
> = 4	224	20.76	1342	16.31

Source Wind, compiled by ZLCA

and commercial credit has become an important financing channel for many companies. The growing accounts receivable takes up an increasing proportion of movable assets. For companies, increasing the recovery rate of accounts receivable is important to reasonably allocate quick ratios.

(3) Cash ratio

Among quick assets, the most liquid one that can be directly used for debt repayment is cash. This chapter uses cash ratio to further measure the short-term debt solvency of Zhongguancun NEEQ companies.

In 2019, two NEEQ companies in Zhongguancun missed data on cash ratio. After excluding them, 328 companies have a cash ratio below 0.5, accounting for 30.40%; 233 companies between 0.5 and 1, accounting for 21.59%; 332 companies between 1 and 3, accounting for 30.77%; 186 companies above 3, accounting for 17.24%. It can be seen that nearly 70% of Zhongguancun NEEQ companies have a cash ratio higher than 0.5, indicating that the overall cash ratio of Zhongguancun NEEQ companies is high and their short-term debt solvency is quite high (see Fig. 4.4).

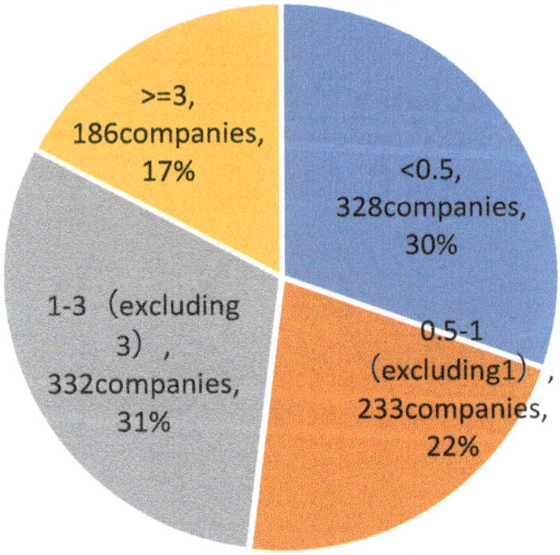

Fig. 4.4 Distribution of cash ratios of Zhongguancun NEEQ companies in 2019. *Source* Wind, compiled by ZLCA

It can be found that the proportion of NEEQ companies with a cash ratio of less than 1 in Zhongguancun is lower than that across the country; while the proportion of companies with a cash ratio higher than 1 is much higher. This shows that compared with the overall cash ratio of all NEEQ companies, the cash ratio of companies in Zhongguancun was generally higher, which means higher ability to use cash to repay debts. On the other hand, this also reveals that some Zhongguancun enterprises have more idle funds (see Table 4.4).

2. Cross-industry Comparison of Short-term Debt Stock Ratio

   Current ratio, quick ratio and cash ratio are relative numbers, and are evidently different in different industries. In order to reduce the impact of the variable of industry, this report calculates the average current ratio, quick ratio and cash ratio of Zhongguancun NEEQ Companies in different industries, and compares these indicators of main industries with the corresponding ones of all NEEQ companies nationwide. Here are the findings:

Table 4.4 Comparison of cash ratio of NEEQ companies across the country and in Zhongguancun in 2019

	Zhongguancun		Nationwide	
	Quantity	Proportion (%)	Quantity	Proportion (%)
<0.5	328	30.40	3555	43.21
0.5–1 (excluding 1)	233	21.59	1927	23.42
1–3 (excluding 3)	332	30.77	1888	22.95
>= 3	186	17.24	858	10.43

Source Wind, compiled by ZLCA

The three indicators of Zhongguancun companies in Scientific Research and Technology, Management of Water Conservancy, Environment and Public Facilities are lower than the national average, and the quick ratio and cash ratio of Information Transmission, Software and Information Technology Service are both lower than the national average, and the quick ratio of Wholesale and Retail Business is lower than the national average; the indicators of Zhongguancun companies in other industries are higher than the national average. To sum up, the overall short-term solvency of NEEQ companies in main industries is higher than that of all NEEQ companies across the nation in the same period (see Table 4.5).

4.2 Analysis of the Operating Capability of Zhongguancun NEEQ Listed Companies

4.2.1 Overview of Operating Capabilities

Operating capability is a financial ratio that measures the efficiency of a company's asset management. It affects the profitability and solvency of modern enterprises. Zhongguancun NEEQ Companies are mainly technological innovation-oriented ones in the growing stage, and their rapid growth does not only rely on breakthroughs in service and products brought by technological research and development but also depend on the efficiency of the company's use of assets. Therefore, the analysis of the operating capability of Zhongguancun NEEQ Listed Companies can help evaluate the growth of a company from the perspective of business efficiency.

Since operating capacity is used to judge the efficiency of asset management, the indicators for measuring operating capacity all reflect how much investment in various assets is required for every 1 yuan of a company's revenue. For enterprises, accounts receivable turnover can reflect the rate of return of revenue, inventory turnover can evaluate a company's inventory management and analyse the inventory occupation of funds; fixed asset turnover reflects the amount of fixed asset investment required for every 1 yuan of revenue generated by the company, reflecting whether the enterprise has the ability to operate with light assets, total asset turnover evaluates the operating capability of an enterprise as a whole. The above indicators are all important for judging the operating capability of an enterprise, and are thus adopted by this report to systematically evaluate the operating capability of Zhongguancun NEEQ Listed Companies from various angles.

4.2.2 Analysis of Accounts Receivable

1. Overall Accounts Receivable

On December 31st, 2019, the total accounts receivable of Zhongguancun NEEQ Companies was 55.724 billion yuan, a decrease of 6.60% from the previous

Table 4.5 Cross-industry comparison of short-term debt stock ratio of NEEQ Companies across the country and in Zhongguancun in 2019

	Average current ratio		Average quick ratio		Average cash ratio	
	Zhongguancun	Nationwide	Zhongguancun	Nationwide	Zhongguancun	Nationwide
Information transmission, software and information technology service	1.96	1.96	1.67	1.69	0.89	0.94
Manufacturing	1.58	1.52	1.15	1.06	0.61	0.55
Leasing and business services	1.27	1.51	1.23	1.45	0.98	1.11
Scientific research and technology service	2.12	1.8	1.7	1.55	0.99	0.8
Culture, sports and entertainment	1.83	0.96	1.15	0.65	0.72	0.42
Wholesale and retail business	1.68	1.49	0.89	0.97	0.56	0.55
Education	2.46	1.31	2.35	1.26	2.14	1.12
Management of Water Conservancy, Environment and Public Facilities	1.39	1.46	1.09	1.14	0.53	0.62

Source Wind, compiled by ZLCA

year. In 2019, the total receivables turnover in days of Zhongguancun NEEQ Companies were 95.17 days, an increase by seven days over 2018. The above data show that although the total amount of accounts receivable of Zhongguancun NEEQ companies has declined, the receivables turnover in days have increased, indicating greater difficulty in collecting money and higher risk of bad debt.
In 2019, the accounts receivable turnover in days of Zhongguancun NEEQ Listed Companies on innovation tier were 129.15 days, an increase of 22.12 days compared with 2018 (107.04 days). This shows: first of all, these companies were in greater pressure on the collection of accounts receivable than the whole Zhongguancun NEEQ companies; secondly, the accounts receivable turnover in days of them have increased significantly compared with the previous year, which indicates declining ability of managing accounts receivable.

2. Distribution of Accounts Receivable in Days

   In 2019, 54 Zhongguancun NEEQ Listed Companies reported accounts receivable turnover in days of fewer than 7 days, accounting for 5.17%, and most of them were in Information Transmission, Software and Information Technology Services. The accounts receivable turnover efficiency of these companies was very high, but there might also be the possibility of affecting revenue due to the tight accounts receivable policy. There were 116 companies with accounts receivable turnover in days between 7 and 30 days, accounting for 11.11%, 261 companies between 30 and 90 days, accounting for 25.00%, 256 companies between 90 and180 days, accounting for 24.52%. The frequency of these companies' accounts receivable turnover was between one quarter Between half a year. There were 357 companies with accounts receivable turnover in days greater than 180 days, accounting for 34.20%. The turnover frequency of these companies was relatively low, and the funds of the enterprises were locked on accounts receivable, which may affect the normal turnover of funds, and may also affect the solvency and financial expenses of the company (see Fig. 4.5).

3. Relationship between Accounts Receivable and Cash and Cash Equivalents

   For accounts receivable, the starting point is a company's credit sales, and the ending point is the cash recovered. Under normal circumstances, the increase in credit sales causes a rise in accounts receivable, and the company's cash stock will also increase. However, if the company's accounts receivable is increasing but its cash is decreasing, the reason may be that credit sales have caused quite serious problems. For example, although relaxed credit policies stimulate customers to make purchase, bad debts increase, and it is difficult to recover cash. In order to more accurately represent the accounts receivable of Zhongguancun NEEQ Listed Companies, this report divides them into four categories from two dimensions accounts receivable and cash and cash equivalents: simultaneous increase in accounts receivable and cash and cash equivalents, increase in accounts receivable and decrease in cash and cash equivalents, decrease in accounts receivable and increase in cash and cash equivalents, and decrease in both accounts receivable and cash and cash equivalents.

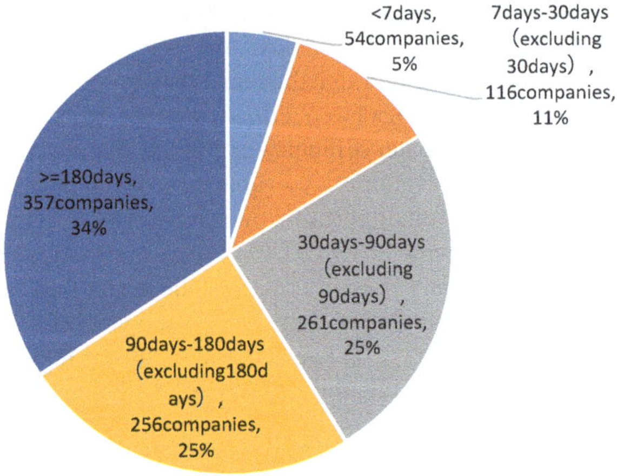

Fig. 4.5 Distribution of accounts receivable turnover in days of Zhongguancun NEEQ listed companies in 2019. *Source* Wind, compiled by ZLCA

Table 4.6 Relationship between accounts receivable and cash and cash equivalents of Zhongguancun NEEQ listed companies in 2019

		Accounts receivable	
		Up	Down
Cash and cash equivalents	Up	240 companies	271 companies
	Down	219 companies	307 companies

Source Wind, compiled by ZLCA

According to the above classification,[3] in 2019, 240 companies (23.14%) reported increase in accounts receivable and cash and cash equivalents, indicating to a certain extent that the growth of these companies' accounts receivable has promoted the increase in their revenue, thereby increasing the cash flow generated from operations; the accounts receivable of 271 companies (26.13%) decreased while their cash and cash equivalents increased, which to a certain extent indicates that these companies have tightened their credit policies, but this has not yet had an impact on the cash flows generated from operations. The accounts receivable of 219 companies (accounting for 21.12%) increased, but their cash and cash equivalents have decreased. The increase in accounts receivable of these companies has not helped the company's cash inflow growth, and there may be a problem that funds cannot be recovered; 307 companies (29.60% of the total) saw decline in both accounts receivable and cash and cash equivalents, and they may have the problem of declining sales scale (see Table 4.6).

[3]Forty-four companies (accounting for 4.07%) did not report accounts receivable in 2018, and were not included in this report.

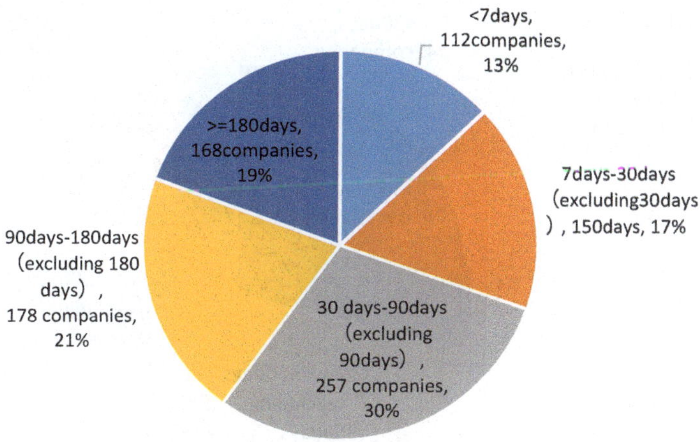

Fig. 4.6 Distribution of inventory turnover in days of Zhongguancun NEEQ listed companies in 2019. *Source* Wind, compiled by ZLCA

4.2.3 Analysis of Inventory Turnover in Days

In 2019, the total inventory turnover in days of Zhongguancun NEEQ Companies were 61.92 days, a decrease of 22.97 days compared with 2018 (84.88 days); the total inventory turnover in days of innovation-tier enterprises were 79.15 days, an increase of 3.42 days compared with 2018 (75.73 days). This shows that the inventory turnover of Zhongguancun NEEQ Listed Companies in 2019 has been significantly improved compared with 2018. The rate of turning inventory into cash through sales has accelerated, and the level of capital occupation by inventory has decreased; while the inventory turnover efficiency of companies on innovation tier has been lower than the overall, and also has dropped.

There were 112 companies whose inventory turnover in days[4] were fewer than 7 days, accounting for 12.95%; 150 companies between 7 and 30 (excluding 30), accounting for 17.34%; 257 companies between 30 and 90 (excluding 90), accounting for 29.71%; 178 companies between 90 and 180 (excluding 180), accounting for 20.58%; 168 companies greater than 180 days, accounting for 19.42% (see Fig. 4.6).

To sum up, in 2019, about one-third of Zhongguancun NEEQ Listed Companies reported inventory turnover in days of fewer than one month, which represents high inventory turnover efficiency and strong inventory management capability. There were also some companies with low inventory turnover efficiency, mainly in manufacturing (61 companies), Information Transmission, Software and Information Technology Service (37 companies), Culture, Sports and Entertainment (30 companies).

---

[4]In 2019, 216 of 1081 Zhongguancun NEEQ Listed Companies reported zero inventory, and are thus excluded in this report for their inventory turnover in days cannot be calculated.

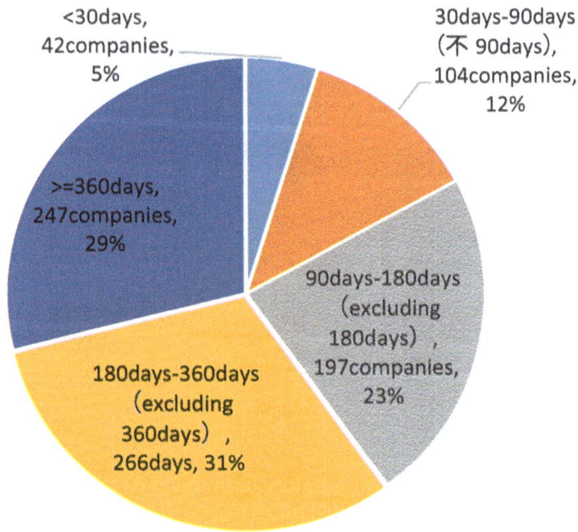

Fig. 4.7 Distribution of operating cycle of Zhongguancun NEEQ listed companies in 2019. *Source* Wind, compiled by ZLCA

4.2.4 Analysis of Operating Cycle

Operating cycle is the sum of inventory turnover in days and accounts receivable turnover in days. It measures the time from the purchase of raw materials before sale to the collection of accounts receivable after sale for every 1 yuan of revenue, representing the operating capability from production to payment.

In 2019,[5] 247 Zhongguancun NEEQ Listed Companies reported an operating cycle of 360 days or more, accounting for 28.55%. It takes close to or more than one year for the inventory obtained by these companies to turn into cash, and the capital turnover rate is slow, which in turn has an impact on the solvency and profitability of the companies; 266 companies had a business cycle between 180 and 360 days (excluding 360 days), accounting for 30.75%. The operating cycle of these companies is between half a year and one year, and thus the capital turnover is quite moderate. There were 197 companies with a business cycle of 90 days to 180 days (excluding 180 days), accounting for 22.77%; 146 companies fewer than 90 days, accounting for 16.99%.The operating cycle of these companies is fewer than a quarter, the capital turnover is fast, and the company's asset management efficiency is relatively high (see Fig. 4.7).

[5]In 2019, 225 of 1081 Zhongguancun NEEQ Listed Companies reported no inventory turnover in days or accounts receivable turnover in days, and are thus excluded in this report for their operating cycle cannot be calculated.

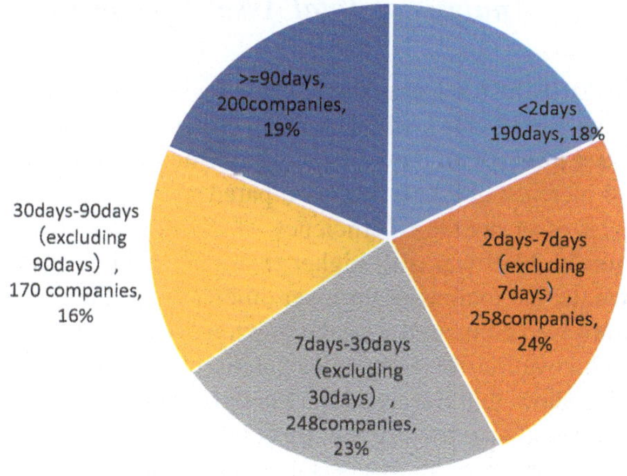

Fig. 4.8 Distribution of fixed asset turnover in days of Zhongguancun NEEQ listed companies in 2019. *Source* Wind, compiled by ZLCA

4.2.5 Analysis of Fixed Assets Turnover in Days

In 2019, the total fixed asset turnover in days of Zhongguancun NEEQ companies were 35.05 days, an increase of 23.97 days compared with 2018 (11.08 days); the total fixed asset turnover in days of innovation-tier companies were 44.04 days, an increase of 32.63 days from 2018 (11.41 days). This shows that the turnover days of fixed assets of innovation-tier enterprises in 2019 are relatively close to that of the whole Zhongguancun NEEQ Listed Companies, and both have increased significantly compared with the previous year.

There were 190 companies with fixed asset turnover in days[6] of fewer than 2 days, accounting for 17.82%; 258 companies between 2 and 7 (excluding 7), accounting for 24.20%; 248 companies between 7 and 30 (excluding 30), accounting for 23.26%; 170 companies between 30 and 90 (excluding 90), accounting for 15.95%; 200 companies more than 90 days, accounting for 18.76%.

On the whole, the number of fixed asset turnover in days of Zhongguancun NEEQ companies in 2019 was relatively small, which was primarily due to two reasons: on the one hand, the utilization efficiency of fixed assets of Zhongguancun NEEQ Companies is generally higher; on the other hand, fixed assets are not the main assets that generate corporate value of Zhongguancun's NEEQ Companies, which allows companies to pay less investment in fixed assets and obtain high revenue (see Fig. 4.8).

[6]In 2018, 15 of 1081 Zhongguancun NEEQ Listed Companies reported no fixed assets, and are thus excluded in this report for their fixed assets turnover in days cannot be calculated.

4.2.6 *Analysis of Total Asset Turnover in Days*

In 2019, the total asset turnover in days of Zhongguancun NEEQ Listed Companies were 666.95 days, which was 8.97 days fewer than in 2018 (675.93 days); the total asset turnover in days of innovation-tier enterprises were 526.84 days, which was an increase of 33.60 days compared with 2018 (493.24 days). This shows that the total asset operating efficiency of Zhongguancun's NEEQ innovation-tier companies in 2019 was also higher than that of Zhongguancun's NEEQ companies as a whole; the total asset turnover efficiency of Zhongguancun's NEEQ companies has improved compared with the previous year, while the total asset operating efficiency of innovation-tier companies has slightly dropped. At the same time, the above data also show that, on average, Zhongguancun NEEQ Listed Companies can recover their total asset investment through sales revenue within 1 to 2 years, which means that their use of all assets is more efficient.

There were 86 companies with total assets turnover in days[7] of fewer than 180 days, accounting for 8.05%; 263 companies between 180 and 360 (excluding 360), accounting for 24.63%; 229 companies between 360 and 540 (excluding 540), accounting for 21.44%; 136 companies between 540 and 720 (excluding 720), accounting for 12.73%; 354 companies 720 days or more, accounting for 33.15%.

The above data show that nearly 70% of Zhongguancun's NEEQ companies can reclaim their total asset investment with revenue within two years. It can be seen that, except some companies showing high asset operating capability, the average asset operating capability of Zhongguancun NEEQ Listed Companies is at the upper-middle level (see Fig. 4.9).

4.3 Conclusion and Recommendation

The above analysis on the solvency of Zhongguancun NEEQ Listed Companies shows: (1) the overall debt ratio of these companies is moderate, and their long-term solvency is strong, but nearly half of them have a low asset-liability ratio, which reflects the ineffective use of financial leverage; (2) cash flow to debt ratio is polarized, with some companies having a weaker ability to repay debts by using cash flow generated from operations, while others having a net cash flow from operations that is much higher than the total corporate debt, and this reflects that these companies have improperly planned for debt, resulting in high financial risks or idle cash assets; (3) in all industries, the current ratio, quick ratio, and cash ratio of Zhongguancun NEEQ companies are generally better than the national average, showing strong short-term solvency.

[7] In 2018, 6 of 1081 Zhongguancun NEEQ Listed Companies (0.82%) reported no revenue in 2018, and thus their fixed assets turnover in days cannot be calculated. According to the data of Wind, the total assets turnover of the six companies was recorded as zero.

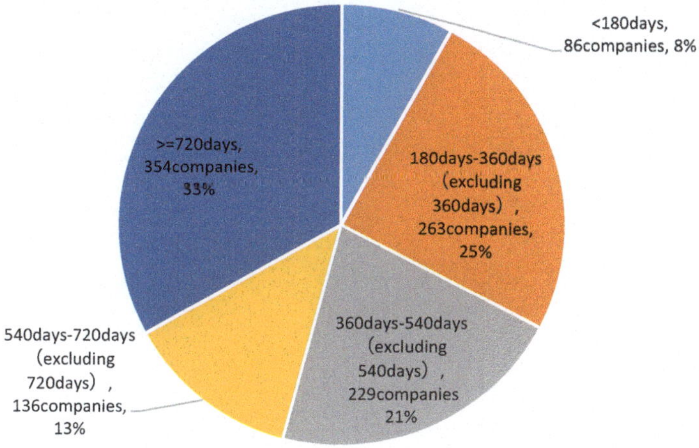

Fig. 4.9 Distribution of total assets turnover in days of Zhongguancun NEEQ listed companies in 2019. *Source* Wind, compiled by ZLCA

The analysis of the operating capability reveals that the total accounts receivable of Zhongguancun NEEQ Listed Companies has declined, but the accounts receivable turnover in days have increased, the efficiency of accounts receivable management has decreased, and the debt collection period of some enterprises has been so long that it may influence their solvency. The operating cycles and fixed asset turnover in days are generally low, highlighting the improvement of the overall management efficiency of Zhongguancun NEEQ companies; most of these companies have a low total asset turnover, and they use only 1–2 years to recover total asset investment in the form of revenue, which represents strong overall operating capability.

Based on the above conclusions and the features of economic development in 2019 and 2020, this report proposes the following recommendations:

(1) Companies should strengthen capital management to ensure the rational and effective use of company funds. At the same time, they should establish a long-term and effective accounts receivable management mechanism to help companies improve the efficiency of capital use. Under the premise of fully considering the interests of shareholders, they should build more funding channels and properly use financial leverage to meet their own necessary capital needs.
(2) Working capital management is related to the solvency and operating capability of a company. The executive team should pay full attention to the management of working capital in financial management. They also should adopt appropriate credit policies, optimize inventory procurement and storage, and rationally use measures like commercial credit financing such as to keep the amount of working capital, the efficiency of account receivable turnover, and the efficiency of inventory turnover at the optimal level, ensure the full liquidity and safety of the company's financial situation, realize the efficient operation of working capital, and improve the company's operating performance.

Chapter 5
The Industrial Distribution Research Report of Zhongguancun NEEQ Listed Companies in 2019

Abstract This paper measures the overall solvency of Zhongguancun NEEQ enterprises through indicators such as asset-liability ratio, cash flow debt ratio, current ratio, etc; and presents the operating capacity of Zhongguancun NEEQ enterprises through indicators such as account receivable turnover days, inventory turnover days, and total asset turnover days. Through corresponding research,we found that the overall debt ratio of Zhongguancun NEEQ enterprises is relatively low and the solvency is good; the fixed assets turnover days, total assets turnover days and other indicators are excellent, and the operating capacity is strong.

Keyword Zhongguancun NEEQ · Industrial distribution · Market performance · Growth

5.1 Overview

According to the "Guidelines on Industry Classification of the Management Type of Listed Companies", this report classifies 1081 Zhongguancun listed companies into 18 sections, 59 divisions, and 156 groups.

5.1.1 Industrial Distribution

1081 Zhongguancun listed companies can be classified into 18 sections, among which the "information transmission, software and IT services" sector takes the lead with 437 listed firms, accounting for 40.43% of the total. "Manufacturing" sector ranks the second with 231 listed companies, amounting to 21.37% of all listed firms. "Leasing and business services" holds a third position with 103 listed companies, representing 9.53% of the all. The rest can be categorized into sectors including "scientific research and technology service", "culture, sports and entertainment", "wholesale and retail businesses", and "education" (Table 5.1). A sector can be further subdivided into 59 divisions, among which Zhongguancun NEEQ listed companies are mainly in the

Zhongguancun Listed Companies Association, *The Growth Report of Zhongguancun NEEQ Listed Companies (2020)*, Current Chinese Economic Report Series,
https://doi.org/10.1007/978-981-33-6819-4_5

Table 5.1 Comparison of the industrial distribution of Zhongguancun NEEQ listed companies between 2019 and 2018

Industrial sector	At the end of 2019		At the end of 2018	
	Number of Enterprises	Proportion (%)	Number of Enterprises	Proportion (%)
Information transmission, software and IT services	437	40.43	485	39.62
Manufacturing	231	21.37	252	20.59
Leasing and business services	103	9.53	129	10.54
Scientific research and technology service	67	6.20	80	6.54
Culture, sports and entertainment	65	6.01	76	6.21
Wholesale and retail businesses	58	5.37	64	5.23
Education	26	2.41	31	2.53
Management of water conservancy, environment and public facilities	21	1.94	22	1.8
Construction	19	1.76	20	1.63
Financials	12	1.11	16	1.31
Transportation, warehousing and postal services	9	0.83	10	0.82
Residential services, repairs and other services	7	0.65	10	0.82
Mining	7	0.65	6	0.49
Health and social work	6	0.56	7	0.57
Electricity, heat, gas and water production and supply industry	6	0.56	5	0.41
Real estate	5	0.46	7	0.57
Agriculture, forestry, animal husbandry and fishery	1	0.09	3	0.25
Accommodation and catering industry	1	0.09	1	0.08

Source Wind, Organized by Zhongguancun Listed Companies Association

"software and IT services", "business service", and "Internet and related services", with 344, 101 and 87 listed companies respectively. A division covers 156 groups, among which the "Software development" group and the "information system integration services" group have the most of Zhongguancun NEEQ listed companies, with 162 and 114 respectively.

According to data, Zhongguancun listed companies are mainly distributed into three industrial sectors: the "information transmission, software and IT services" represented by "software and IT services" division and "Internet and related services"

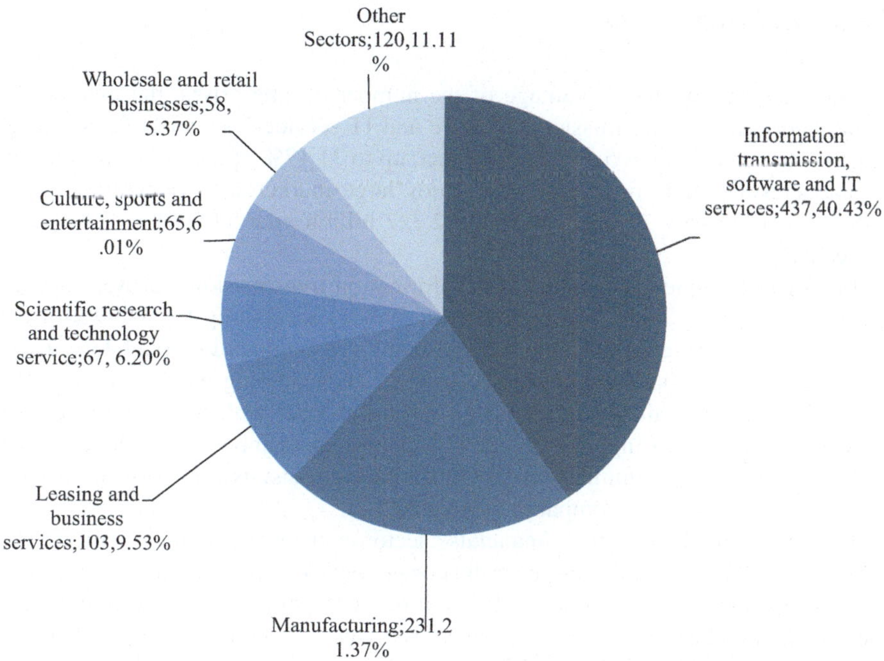

Fig. 5.1 Industrial Distribution of Zhongguancun NEEQ Listed Companies in 2019. *Note 1* "Other sectors" cover education (26, 2.41%); management of water conservancy, environment and public facilities (21, 1.94%); construction (19, 1.76%); financials (12, 1.11%); transportation, warehousing and postal services (9, 0.83%); residential services, repairs and other services (7, 0.65%); mining (7, 0.65%); health and social work (6, 0.56%); electricity, heat, gas and water production and supply industry (6, 0.56%); real estate (5, 0.46%); agriculture, forestry, animal husbandry and fishery (1, 1.09%); and accommodation and catering industry ((1, 1.09%)
Note 2 Percentages do no equal 100% as the result displays up to two decimal places
Source Wind, Organized by Zhongguancun Listed Companies Association

division, the "manufacturing" with "computer, telecommunications, and other electronic equipment manufacturing" and "special equipment manufacturing" as representatives, and the "leasing and business services" represented by the "business services" division. Except for those three sectors, "scientific research and technology service", "culture, sports and entertainment", and "wholesale and retail businesses" also share a large number of Zhongguancun listed companies (see Fig. 5.1).

In general, Zhongguancun has attracted nearly 30,000 high-tech enterprises for over 30 years, and has formed a high and new-tech industrial cluster featuring strong R&D and services. We are pleased to see that Zhongguancun has become one of the most intensive areas in China for talent resources and science and education intelligence.

5.1.2 Industrial Scale

Considering the absolute advantage of the number of enterprises, the market cap of the "information transmission, software and IT services" sector is far ahead of other sectors, reaching 105.970 billion yuan, up to 31.17% of the total. In addition, "manufacturing" and "financials" also enjoy large market caps, generating 64.44 billion yuan (18.95% of the total) and 52.227 billion yuan (15.38% of the total) respectively.

In terms of employee numbers, the"information transmission, software and IT services" sector has the most employees of 59,826, accounting for 34.99% of the total. "Manufacturing" contributes 41, 711 employees, amounting to 24.43% of the total. Other sectors have employees below 10%. Considering the average number of employees,[1] the "financials" division takes the lead, with 400 employees per company. "Manufacturing" sector has higher average number of employees (181 per company) than the number of the "information transmission, software and IT services" sector (137 per company) (see Table 5.2).

In terms of total asset, the "financials" sector contributes the most with 95.108 billion yuan (25.83%) due to its generally considerable enterprise scales.[2] Its average asset is also the largest, reaching 7.926 billion yuan per company, far higher than the average number. "Leasing and business services", "manufacturing", and "information transmission, software and IT services" sectors also enjoy high total assets, reaching 19.95%, 18.81%, and 18.56% respectively. Other sectors share less than 5% total assets. Net asset has a similar feature. The top three sectors are, by net asset, "financials", "manufacturing", and "information transmission, software and IT services" sectors, accounting for 28.46%, 21.54% and 20.24% respectively of the total (see Table 5.3).

Considering market cap, asset, and employee scale, "information transmission, software and IT services", "manufacturing", and "leasing and business services" sectors have large overall scales as they have an absolute advantage in enterprise number. "Financials" have less enterprises but have better performances than other sectors in terms of average market cap and average asset. This is because the financial sector has special industrial features.

[1]This report calculates but does not go details into industrial sectors with enterprises below 1% of the total, including "transportation, warehousing and postal services" (9, 0.83%), "residential services, repairs and other services" (7, 0.65%), "mining" (7, 0.65%), "health and social work" (6, 0.56%), "electricity, heat, gas and water production and supply industry" (6, 0.56%), "real estate" (5, 0.46%), "agriculture, forestry, animal husbandry and fishery" (1, 1.09%), and "accommodation and catering industry" (1, 1.09%) for industrial key indicators.

[2]12 enterprises belong to the "financials" sector. Among all Zhongguancun NEEQ listed companies, Jiuding Group (430,719.OC) takes the lead with 46.263 billion-yuan total assets. Guodu Securities (870,488.OC) has a total asset of 29.208 billion yuan. Guodu Securities (870,488.OC), China Equity Group (833,858.OC), Heaven-sent Capital Management Group (833,044.OC), China Post Fund (834,344.OC), and NewMargin Venture Capital (833,502.OC) have total assets exceeding 1 billion yuan.

Table 5.2 Market cap and employee scale of Zhongguancun NEEQ listed companies in 2019

Industrial sector	Market cap			Employee number		
	The total	Proportion (%)	The average	The total	Proportion (%)	The average
Information transmission, software and IT services	1059.70	31.17	2.42	59,826	34.99	137
Manufacturing	644.40	18.95	2.79	41,771	24.43	181
Financials	522.77	15.38	43.56	4795	2.80	400
Leasing and business services	323.61	9.52	3.14	9953	5.82	97
Culture, sports and entertainment	201.21	5.92	3.10	4350	2.54	67
Scientific research and technology service	194.66	5.73	2.91	13,160	7.70	196
Wholesale and retail businesses	174.05	5.12	3.00	6421	3.76	111
Construction	66.92	1.97	3.52	3087	1.81	162
Management of water conservancy, environment and public facilities	61.50	1.81	2.93	4022	2.35	192
Education	58.07	1.71	2.23	3916	2.29	151
Transportation, warehousing and postal services	28.67	0.84	3.19	1529	0.89	170
Health and social work	23.14	0.68	3.86	2752	1.61	459
Residential services, repairs and other services	13.81	0.41	1.97	657	0.38	94
Electricity, heat, gas and water production and supply industry	12.10	0.36	2.02	789	0.46	132
Real estate	8.49	0.25	1.70	13,467	7.88	2693
Mining	3.10	0.09	0.44	352	0.21	50
Accommodation and catering industry	2.00	0.06	2.00	59	0.03	59
Agriculture, forestry, animal husbandry and fishery	1.86	0.05	1.86	76	0.04	76

Source Wind, Organized by Zhongguancun Listed Companies Association

Table 5.3 Assets of Zhongguancun NEEQ listed companies in 2019

Industrial sector	Total assets			Net assets		
	The total	Proportion (%)	The average	The total	Proportion (%)	The average
Financials	951.08	25.83	79.26	536.96	28.46	44.75
Leasing and business services	734.58	19.95	7.13	247.16	13.10	2.40
Manufacturing	692.69	18.81	3.00	406.30	21.54	1.76
Information transmission, software and IT services	683.58	18.56	1.56	381.85	20.24	0.87
Scientific research and technology service	129.14	3.51	1.93	79.55	4.22	1.19
Wholesale and retail businesses	125.99	3.42	2.17	55.13	2.92	0.95
Culture, sports and entertainment	117.76	3.20	1.81	57.30	3.04	0.88
Construction	68.31	1.86	3.60	23.88	1.27	1.26
Management of water conservancy, environment and public facilities	63.11	1.71	3.01	29.98	1.59	1.43
Transportation, warehousing and postal services	41.97	1.14	4.66	29.53	1.57	3.28
Electricity, heat, gas and water production and supply industry	21.48	0.58	3.58	7.89	0.42	1.31
Education	21.18	0.58	0.81	14.58	0.77	0.56
Real estate	17.22	0.47	3.44	8.70	0.46	1.74
Health and social work	7.14	0.19	1.19	4.27	0.23	0.71
Mining	2.88	0.08	0.41	1.70	0.09	0.24
Residential services, repairs and other services	2.57	0.07	0.37	1.35	0.07	0.19
Agriculture, forestry, animal husbandry and fishery	1.35	0.04	1.35	0.27	0.01	0.27
Accommodation and catering industry	0.22	0.01	0.22	0.12	0.01	0.12

Source Wind, Organized by Zhongguancun Listed Companies Association
100 million yuan

5.2 Comparative Analysis of Industrial Profitability

"Information transmission, software and IT services" sector is the largest by overall revenue, reaching 76.718 billion yuan (37.12% of the total), due to its absolute advantage in the enterprise number. "Manufacturing", "wholesale and retail businesses", and "leasing and business services" also enjoy high revenues, up to 39.658 billion yuan, 35.187 billion yuan, and 20.368 billion yuan respectively. Other sectors have revenues less than 10 billion yuan. In terms of average revenue, "wholesale and retail businesses" and "financials" sectors perform better than other sectors with 607 million yuan and 586 million yuan average revenues respectively (Table 5.4).

Table 5.4 Revenues of Zhongguancun NEEQ listed companies in 2019

Industrial Sector	Revenue (100 million yuan)	Average Revenue (100 million yuan)	Maximum Revenue (100 million yuan)	Revenue per Capita (10 thousand yuan/person)	Proportion (%)
Information transmission, software and IT services	767.18	1.76	166.02	128.23	37.12
Manufacturing	396.58	1.72	52.95	94.94	19.19
Wholesale and retail businesses	351.87	6.07	208.36	548.00	17.02
Leasing and business services	203.68	1.98	26.71	204.64	9.85
Scientific research and technology service	72.27	1.08	8.85	54.91	3.50
Financials	70.27	5.86	23.93	146.54	3.40
Construction	46.32	2.44	13.82	150.05	2.24
Culture, sports and entertainment	45.92	0.71	6.58	105.57	2.22
Management of water conservancy, environment and public facilities	30.12	1.43	4.84	74.88	1.46
Transportation, warehousing and postal services	21.50	2.39	9.39	140.63	1.04
Education	19.73	0.76	6.25	50.38	0.95
Real estate	19.18	3.84	10.31	14.24	0.93
Health and social work	9.80	1.63	5.18	35.63	0.47
Electricity, heat, gas and water production and supply industry	6.22	1.04	1.93	78.85	0.30
Residential services, repairs and other services	3.62	0.52	1.08	55.02	0.17
Mining	1.61	0.23	0.53	45.87	0.08
Agriculture, forestry, animal husbandry and fishery	0.77	0.77	0.77	100.86	0.04
Accommodation and catering industry	0.20	0.20	0.20	33.13	0.01

Source Wind, Organized by Zhongguancun Listed Companies Association

"Information transmission, software and IT services" and "manufacturing" have reasonable total gross profits considering the number of enterprises in each sector, reaching 16.514 billion yuan (35.41%) and 12.792 billion yuan (27.43%) respectively. Those numbers are far greater than other sectors. The "financials" sector tops 280 million yuan in average gross profit, while the average gross profit of other sectors are below 100 million yuan. This indicates that the "financials" sector enjoys high gross profit.

"Financials" and "Education" rank first and second among all sectors for average gross profits, reaching 47.79% and 40.63% respectively. A total number of nine sectors, including "health and social work", "mining", "scientific research and technology service", "manufacturing", "electricity, heat, gas and water production and supply industry", "residential services, repairs and other services", "agriculture, forestry, animal husbandry and fishery", "culture, sports and entertainment", and "management of water conservancy, environment and public facilities", have average gross profits ranging from 30.00% to 40.00% (see Table 5.5).

Table 5.5 Gross profits of Zhongguancun NEEQ listed companies in 2019

Industrial sector	Total gross profit (100 million yuan)	Maximum gross profit (100 million yuan)	Average gross profit (100 million yuan)	Average gross margin (%)	Proportion (%)
Information transmission, software and IT services	165.14	13.37	0.38	21.53	35.41
Manufacturing	127.92	12.98	0.55	32.25	27.43
Leasing and business services	37.69	6.74	0.37	18.51	8.08
Financials	33.58	16.23	2.80	47.79	7.20
Wholesale and retail businesses	27.43	7.48	0.47	7.80	5.88
Scientific research and technology service	23.90	2.93	0.36	33.08	5.13
Culture, sports and entertainment	13.84	2.89	0.21	30.15	2.97
Management of water conservancy, environment and public facilities	9.05	1.58	0.43	30.04	1.94
Education	8.02	1.23	0.31	40.63	1.72
Construction	6.85	1.83	0.36	14.78	1.47
Health and social work	3.84	2.10	0.64	39.19	0.82
Real estate	2.84	1.13	0.57	14.83	0.61
Transportation, warehousing and postal services	2.20	1.02	0.24	10.25	0.47
Production and supply of electricity, heat, gas and water	2.00	0.98	0.33	32.18	0.43
Electricity, heat, gas and water production and supply industry	1.13	0.24	0.16	31.33	0.24
Mining	0.57	0.20	0.08	35.41	0.12
Agriculture, forestry, animal husbandry and fishery	0.24	0.24	0.24	31.21	0.05
Accommodation and Catering	0.05	0.05	0.05	25.69	0.01

Source Wind, Organized by Zhongguancun Listed Companies Association

Table 5.6 Net profits of Zhongguancun NEEQ listed companies in 2019

Industrial sector	Total net profit (100 million yuan)	Maximum net profit (100 million yuan)	Average net profit (100 million yuan)	Average Net Margin (%)	Net Profit per Capita (10 thousand yuan)
Financials	28.07	19.48	2.34	27.72	58.54
Manufacturing	25.47	5.67	0.11	1.43	6.10
Leasing and business services	13.12	8.31	0.13	4.08	13.18
Information transmission, software and IT services	12.27	6.21	0.03	0.81	2.05
Scientific research and technology service	4.87	1.20	0.07	1.66	3.70
Wholesale and retail businesses	2.77	2.16	0.05	0.61	4.31
Management of water conservancy, environment and public facilities	1.56	0.41	0.07	1.35	3.88
Real estate	0.78	0.28	0.16	1.45	0.58
Health and social work	0.30	0.21	0.05	2.15	1.09
Education	0.23	0.40	0.01	2.02	0.59
Agriculture, forestry, animal husbandry and fishery	0.08	0.08	0.08	10.76	10.85
Accommodation and catering	0.01	0.01	0.01	5.53	1.83
Residential services, repairs and other services	0.03	0.04	0.00	1.03	0.51
Mining	0.10	0.05	0.01	3.09	2.73
Electricity, heat, gas and water production and supply industry	0.14	0.60	0.02	9.60	1.75
Transportation, warehousing and postal services	1.52	0.21	0.17	0.99	9.93
Construction	2.77	0.43	0.15	0.93	8.97
Culture, sports and entertainment	3.60	1.11	0.06	2.42	8.27

Source Wind, Organized by Zhongguancun Listed Companies Association

"Financials" and "manufacturing" sectors hold total net profits of 2.807 billion yuan and 2,547 billion yuan respectively twice that of the "leasing and business services" (1,312 billion yuan) and "information transmission, software and IT services" (1,227 billion yuan). In particular, the "financials" sector far outpaces other sector for average net profits (234 million yuan), average net margin (27.72%), and net profit per capita (585,400 yuan) (see Table 5.6).

Statistics indicate Zhongguancun NEEQ companies listed on innovation tier are mainly from 7 industrial sectors, among which the "information transmission, software and IT services" sector (45) and the "manufacturing" sector (19), in particular, contribute up to 71.11% of the total, as those sectors have a sizable number of enterprises. In addition, the average net profit generated by innovation tier companies is generally higher than that of base tier companies, consistent with the fact that the overall average net profit of innovation tier enterprises (21 million yuan) is higher than that of base tier enterprises (600 million yuan) (see Table 5.7).

The "agriculture, forestry, animal husbandry and fishery" sector arrives at the highest ROA of 6.00%. In addition, the ROA of innovation tier enterprises (2.80%)

Table 5.7 Profitability of Zhongguancun NEEQ Companies listed on base tier and innovation tier

Industrial sector	Base				Innovation			
	No. of NEEQ listed companies	Net profit (100 million yuan)	Proportion / %	Average net profit (100 million yuan)	No. of NEEQ listed companies	Net profit (100 million yuan)	Proportion/%	Average net profit (100 million yuan)
Financials	12	28.07	45.22	2.34	–	–	–	–
Leasing and business services	97	13.46	21.68	0.14	6	0.34	1.74	0.06
Manufacturing	212	10.56	17.02	0.05	19	14.90	77.24	0.78
Information transmission, software and IT services	392	7.65	12.32	0.02	45	4.62	23.97	0.10
Scientific research and technology service	61	3.81	6.14	0.06	6	1.06	5.48	0.18
Wholesale and retail businesses	56	2.42	3.90	0.04	2	0.35	1.79	0.17
Management of water conservancy, environment and public facilities	20	1.27	2.05	0.06	1	0.29	1.51	0.29
Real estate	5	0.78	1.26	0.16	–	–	–	–
Health and social work	6	0.30	0.48	0.05	–	–	–	–
Agriculture, forestry, animal husbandry and fishery	1	0.08	0.13	0.08	–	–	–	–
Education	25	0.07	0.11	0.00	1	0.16	0.84	0.16

(continued)

Table 5.7 (continued)

Industrial sector	Base				Innovation			
	No. of NEEQ listed companies	Net profit (100 million yuan)	Proportion / %	Average net profit (100 million yuan)	No. of NEEQ listed companies	Net profit (100 million yuan)	Proportion/%	Average net profit (100 million yuan)
Accommodation and catering	1	0.01	0.02	0.01	–	–	–	–
Residential services, repairs and other services	7	0.03	0.05	0.00	–	–	–	–
Mining	7	0.10	0.16	0.01	–	–	–	–
Production and supply of electricity, heat, gas and water	6	0.14	0.22	0.02	–	–	–	–
Transportation, warehousing and postal services	8	0.66	1.06	0.08	1	0.86	4.47	0.86
Culture, sports and entertainment	57	2.67	4.30	0.05	8	0.93	4.80	0.12
Construction	18	2.80	4.51	0.16	1	0.03	0.18	0.03
Overall	991	62.08	99.99[a]	0.06	90	19.29	100.00	0.21

Source Wind, Organized by Zhongguancun Listed Companies Association
"–" represents there is no innovation tier enterprise in this industrial sector
[a]Percentages do no equal 100% as the result displays up to two decimal places

Table 5.8 ROA of Zhongguancun NEEQ listed companies in 2019

Industrial sector	Return on assets (%)		
	Innovation	Base	Overall
Agriculture, forestry, animal husbandry and fishery	–	6.00	6.00
Real estate	–	4.91	4.91
Health and social work	–	4.26	4.26
Scientific research and technology service	5.77	3.57	3.89
Manufacturing	4.91	2.84	3.77
Management of water conservancy, environment and public facilities	6.13	2.28	2.58
Financials	–	2.48	2.48
Wholesale and retail businesses	4.47	2.06	2.21
Leasing and business services	1.20	2.00	1.88
Information transmission, software and IT services	1.88	1.84	1.86
Education	25.49	0.36	1.17
Production and supply of electricity, heat, gas and water	–	0.63	0.63
Residential services, repairs and other services	–	1.39	1.39
Culture, sports and entertainment	1.75	4.11	3.04
Mining	–	3.15	3.15
Transportation, warehousing and postal services	3.18	4.45	3.63
Construction	3.17	4.25	4.13
Accommodation and catering	–	4.58	4.58
Overall	2.80	2.01	2.15

Source Wind, Organized by Zhongguancun Listed Companies Association
"–" indicates there is no innovation tier enterprise in this industrial sector

is generally higher than that of base tier enterprises (2.01%). Among industrial sectors both have companies listed on innovation tier and base tier, the "scientific research and technology service" sector tops with 3.89% of the total (see Table 5.8).

In general, the high profit "financials" sector performs well in revenue, gross profit, net profit, gross margin, and net margin. Other sectors including "Information transmission, software and IT services", "manufacturing", "wholesale and retail businesses", "education", and "leasing and business services" also perform well. In terms of average net profit and ROA, innovation tier enterprises perform better than base tier enterprises.

5.3 Comparative Analysis of Industrial Operating Capability

The "accommodation and catering" sector holds a much higher inventory turnover ratio (504.88) than any other sector, meaning that the sector operates efficiently as fast inventory turnover occurs. The "agriculture, forestry, animal husbandry and fishery" sector and the "health and social work" sector have an accounts receivable turnover ratio of 30.98 and 26.15 respectively, which indicates that customers pay fast and enterprises have strong short-term solvency. Considering total assets turnover, the "wholesale and retail businesses" sector ranks the highest of 2.8, indicating enterprises in this sector take fewer days for cash replenishment and have stronger operating capabilities than others (see Table 5.9).

Table 5.9 Operation capability of Zhongguancun NEEQ listed companies in 2019

Industrial sector	Inventory turnover (Times)	Accounts receivable turnover (Times)	Total assets turnover (Times)
Accommodation and catering	504.88	3.98	0.83
Real estate	38.31	9.01	1.20
Leasing and business services	21.55	4.71	0.29
Health and social work	17.14	26.15	1.39
Education	14.92	15.45	1.00
Transportation, warehousing and postal services	9.65	2.73	0.51
Information transmission, software and IT services	7.73	3.51	1.16
Wholesale and retail businesses	5.99	16.70	2.80
Residential services, repairs and other services	4.78	9.48	1.51
Scientific research and technology service	3.02	2.62	0.58
Mining	2.85	1.55	0.53
Manufacturing	2.54	3.00	0.59
Agriculture, forestry, animal husbandry and fishery	2.39	30.98	0.56
Construction	2.33	2.13	0.69
Management of water conservancy, environment and public facilities	2.24	1.67	0.50
Production and supply of electricity, heat, gas and water	1.80	2.35	0.29
Culture, sports and entertainment	0.86	1.69	0.39
The financial sector	0.85	5.25	0.06

Source Wind, Organized by Zhongguancun Listed Companies Association

5.4 Comparative Analysis of Industrial Solvency

The asset-liability structure of innovation-tier enterprises, with a 42.60% asset-liability ratio, is better than that of base-tier companies and the average; The overall asset-liability ratio of Zhongguancun NEEQ listed companies is 48.77%, and that of base-tier enterprises is 50.26%. From the perspective of industrial sectors, traditional industrial sectors have higher liability ratio compared with emerging high-tech industries; asset-heavy industries have higher liability ratio than asset-light industries. Traditional industries and asset-heavy industries, relatively speaking, have scale advantages, so it is easier to obtain debt financing, but their development prospects are limited, so it is more difficult to obtain equity financing. High-tech industries and asset-light industries, in comparison, are relatively risky in the early stage but offer relatively bright prospects; therefore, it is more difficult to obtain debt financing but easier to obtain equity financing (see Table 5.10).

Table 5.10 Asset-liability ratio of Zhongguancun NEEQ listed companies in 2019

Industrial sector	Asset-liability ratio (%)		
	Base	Innovation	Overall
Agriculture, forestry, animal husbandry and fishery	80.03	–	80.03
Leasing and business services	67.63	34.59	66.35
Construction	65.23	52.85	65.04
Production and supply of electricity, heat, gas and water	63.27	–	63.27
Culture, sports and entertainment	58.14	43.21	51.34
Wholesale and retail businesses	57.52	38.07	56.24
Management of water conservancy, environment and public facilities	52.94	47.29	52.49
Real estate	49.49	–	49.49
Transportation, warehousing and postal services	47.64	19.42	29.64
Residential services, repairs and other services	47.56	–	47.56
Accommodation and catering	45.13	–	45.13
Financials	43.54	–	43.54
Manufacturing	43.16	39.08	41.34
Mining	41.10	–	41.10
Health and social work	40.27	–	40.27
Information transmission, software and IT services	40.09	50.55	44.14
Scientific research and technology service	39.50	32.26	38.40
Education	30.68	44.27	31.15
Overall	50.26	42.60	48.77

Source Wind, Organized by Zhongguancun Listed Companies Association
"–" indicates there is no innovation tier enterprise in this industrial sector

In general, the "transportation, warehousing and postal services" sector ranks highest in current ratio (2.72) and quick ratio (2.52) in 2019, followed by the "education" sector with 2.46 and 2.35 respectively. Traditional industries, such as "electricity, heat, gas and water production and supply industry" and "agriculture, forestry, animal husbandry and fishery", have relatively low current ratio and quick ratio for their relatively weak short-term solvency. In most industries, the quick ratio is around 1, which is within a reasonable range. Companies relying on cash flows record, such as "wholesale and retail businesses", almost no accounts receivable, and have a reasonable quick ratio below 1 (see Table 5.11).

Table 5.11 Current ratio and quick ratio of Zhongguancun NEEQ listed companies by sector in 2019

Industrial sector	Current ratio			Quick ratio		
	Base	Innovation	Overall	Base	Innovation	Overall
Transportation, warehousing and postal services	1.34	4.28	2.72	1.22	4.00	2.52
Education	2.47	2.23	2.46	2.35	2.23	2.35
Scientific research and technology service	2.04	2.68	2.12	1.62	2.27	1.70
Information transmission, software and IT services	2.10	1.78	1.96	1.86	1.43	1.67
Residential services, repairs and other services	1.86	–	1.86	1.39	–	1.39
Culture, sports and entertainment	1.55	2.28	1.83	1.00	1.40	1.15
Financials	1.69	–	1.69	1.59	–	1.59
Wholesale and retail businesses	1.64	2.58	1.68	0.85	1.80	0.89
Accommodation and catering	1.62	–	1.62	1.61	–	1.61
Mining	1.61	–	1.61	1.32	–	1.32
Health and social work	1.59	–	1.59	1.43	–	1.43
Manufacturing	1.77	1.29	1.58	1.31	0.89	1.15
Real estate	1.45	–	1.45	1.39	–	1.39
Management of water conservancy, environment and public facilities	1.33	2.24	1.39	1.07	1.37	1.09
Leasing and business services	1.22	2.18	1.27	1.18	2.02	1.23
Construction	1.23	1.83	1.24	0.83	1.27	0.84
Production and supply of electricity, heat, gas and water	0.80	–	0.80	0.55	–	0.55
Agriculture, forestry, animal husbandry and fishery	0.40	–	0.40	0.20	–	0.20
Overall	1.65	1.74	1.67	1.37	1.33	1.36

Source Wind, Organized by Zhongguancun Listed Companies Association
"–" indicates there is no innovation tier enterprise in this industrial sector

5.5 Comparative Analysis of Industrial R&D Capability

The "information transmission, software and IT services" sector spends the most on R&D[3] with 4.59 billion yuan (54.32% of the total), much higher than the combined total R&D expenses of the "manufacturing" sector, ranking second with 2.478 billion yuan (29.32% of the total), and that of any other industrial sector. In addition, the "information transmission, software and IT services" sector has an average R&D expense of 12 million yuan, ranking second and following the "financials" sector with the highest average R&D expense of 31 million yuan.[4] Regarding R&D intensity, the "manufacturing" sector and the "information transmission, software and IT services" sector are fairly similar with 6.31% and 6.06% respectively, lower than the "mining" sector (see Table 5.12).

Table 5.12 Industrial profitability of companies listed on NEEQ base tier and innovation tier

Industrial sector	R&D expenses (100 million yuan)	Average R&D expenses (100 million yuan)	Proportion (%)	R&D intensity (%)
Information transmission, software and IT services	45.90	0.12	54.32	6.06
Manufacturing	24.78	0.11	29.32	6.31
Scientific research and technology service	3.69	0.07	4.36	5.33
Leasing and business services	2.27	0.04	2.68	2.15
Management of water conservancy, environment and public facilities	1.45	0.07	1.71	4.80
Construction	1.39	0.10	1.65	3.54
Financials	1.24	0.31	1.47	2.95
Culture, sports and entertainment	1.21	0.04	1.43	4.68
Wholesale and retail businesses	0.98	0.03	1.16	0.35
Education	0.64	0.04	0.76	3.81

(continued)

[3]862 out of 1081 Zhongguancun NEEQ listed companies expose R&D expenses.

[4]4 enterprises in the "financials" sector, including China Equity Group (833,858.OC), Bocheng Insurance (870,032.OC), Jiuding Group (430,719.OC), and Unitedmoney (872,088.OC) expose their R&D expenses with 72 million yuan, 23 million yuan, 23 million yuan, 5 million yuan respectively. As R&D expenses are relatively high in those four companies and a few financial companies expose their R&D expenses, the average number is considerable. If we solely take R&D expenses of top four enterprises, ElexTech (872,801.OC), Sincetimes Network(872,801.OC), Yonyou Fintech (839,483.OC), and Hisign Technology (430,021.OC) with the R&D expenses of 378 million yuan, 151 million yuan, 135 million yuan, and 124 million yuan prespectively, into consideration, the average R&D expenses of the "information transmission, software and IT services" sector reaches 197 million yuan, higher than the average number of the "financials" sector.

Table 5.12 (continued)

Industrial sector	R&D expenses (100 million yuan)	Average R&D expenses (100 million yuan)	Proportion (%)	R&D intensity (%)
Transportation, warehousing and postal services	0.35	0.09	0.41	2.55
Production and supply of electricity, heat, gas and water	0.21	0.04	0.25	4.46
Real estate	0.16	0.05	0.19	0.96
Mining	0.12	0.02	0.14	9.28
Residential services, repairs and other services	0.11	0.03	0.13	5.35
Health and social work	0.02	0.02	0.03	4.35
Agriculture, forestry, animal husbandry and fishery	–	–	–	–
Accommodation and catering	–	–	–	–
Overall	84.50	0.10	100.00	0.05

"–" indicates this industrial sector does not expose its R&D expenses

In general, the "information transmission, software and IT services" sector and the "manufacturing" sector have strong R&D capabilities, which is partly because they have a large number of enterprises. This also indicates that the two sectors attach great importance to R&D capabilities.

Chapter 6
Comparative Analysis of the Development of NEEQ Listed Companies in Major Regions in 2019

Abstract Through content comparison, this report analyzes the five regions with most concentrated distribution of NEEQ, including Guangdong, Zhongguancun, Jiangsu, Zhejiang and Shanghai by core indicators such as the market value, assets, revenue, profit. This report also focus on analyzing the overall situation,profitability and innovation ability of each region to highlight the actual development situation and growth ability. According to statistics, affected by the overall economic situation, the development of NEEQ enterprises in five regions in 2019 is not very optimistic, and the number of delisted companies is relatively large, but the growth momentum of NEEQ enterprises that continue to operate in major regions from 2018 to 2019 is still strong. The overall quality and growth of Zhongguancun NEEQ enterprises are still better than other regions, and their innovation ability is relatively strong. Especially after the introduction of the NEEQ selective layer reform, Zhongguancun NEEQ enterprises are actively preparing for war, and many indicators are leading in the country.

Keyword NEEQ · Regional analysis · Innovation ability

6.1 Comparative Analysis of the Overall Situation in Each Region

As of the end of 2019, the total number of NEEQ listed companies was 8,953, a year-on-year decrease of 16.26% compared with 10,691 at the end of 2018. The top five regions with the largest number of NEEQ listed companies were Guangdong (1319), Zhongguancun (1190), Jiangsu (1072), Zhejiang (786) and Shanghai (715), all representing a significant decline from 2018. The number of NEEQ listed companies in the above five regions in 2019 was 5,082, accounting for 56.76% of the total number over the country. (See Table 6.1).

By June 30, 2020, the number of NEEQ companies that have published their annual reports were 1200 in Guangdong, 1081 in Zhongguancun, 984 in Jiangsu, 730 in Zhejiang and 669 in Shanghai. As this report involves the analysis of 2019 financial data, the enterprises that published their annual reports before June 30, 2020

Zhongguancun Listed Companies Association, *The Growth Report of Zhongguancun NEEQ Listed Companies (2020)*, Current Chinese Economic Report Series,
https://doi.org/10.1007/978-981-33-6819-4_6

Table 6.1 Number of NEEQ companies in major regions

Regions	The end of 2019		The end of 2018		Growth rate (%)		Number of Going Concerns from 2018 to 2019
	Number of companies	Proportion (%)	Number of companies	Proportion (%)	Number of companies	Proportion	
Guangdong	1319	14.73	1637	15.31	−19.43	−3.79	1136
Zhongguancun	1190	13.29	1440	13.47	−17.36	−1.34	1035
Jiangsu	1072	11.97	1273	11.91	−15.79	0.50	927
Zhejiang	786	8.78	933	8.73	−15.76	0.57	689
Shanghai	715	7.99	903	8.45	−20.82	−5.44	649

Source Wind, compiled by ZLCA

will be selected as the research objects in the following analysis of key indicators in each region.

6.1.1 Market Capitalization

In 2019, the market cap of NEEQ listed companies in Zhongguancun was 340.005 billion yuan. Correspondingly, the market cap in Guangdong, Jiangsu, Zhejiang and Shanghai were 316.388 billion yuan, 217.023 billion yuan, 184.549 billion yuan and 216.884 billion yuan. On the whole, the market cap of the NEEQ listed companies in all the above regions has shown a downward trend over the past two years, but the market cap of Zhongguancun NEEQ listed companies werestill ahead of other regions (see Fig. 6.1).

Judging from the market cap of the NEEQ listed companies that have operated for two consecutive years from 2018 to 2019, the market cap of continuously operating companies in Zhongguancun and Jiangsu have basically remained unchanged compared with last year. The market cap of continuously operating companies in Guangdong has increased by about 5%, while the market cap of Zhejiang and Shanghai have increased significantly, with a growth rate above 10%.

Considering the market cap of the leading companies, there were, at the end of 2019, three Zhongguancun NEEQ listed companies with a market cap of over 10 billion yuan, among which Jiuding Group had a highest market cap of 11.55 billion yuan. There were three in Guangdong, one in Zhejiang and one in Shanghai, which shared a market cap of more than 10 billion yuan respectively. The market cap of the NEEQ listed companies in Jiangsu were all below 10 billion yuan.

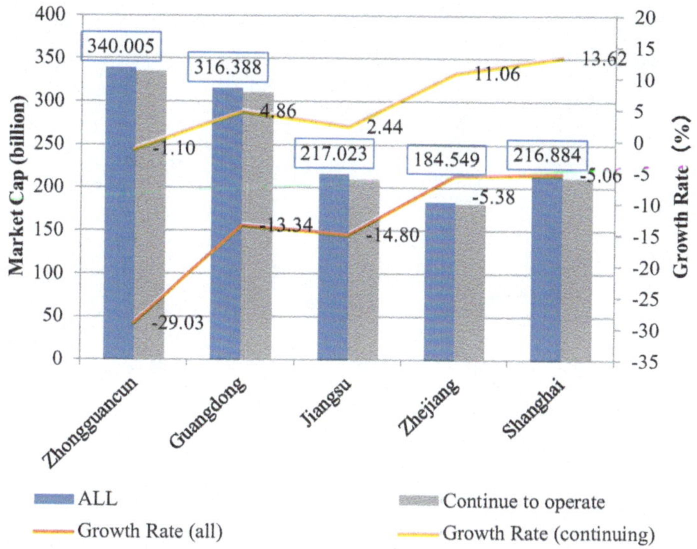

Fig. 6.1 The market cap of NEEQ listed companies in each region at the end of 2019. *Source* Wind, compiled by ZLCA

6.1.2 Assets

In 2019, the total assets of Zhongguancun NEEQ listed companies were 368.225 billion yuan, ranking first. NEEQ listed companies in Jiangsu ranked the second, with total assets of 314.329 billion yuan. The total assets of Guangdong, Zhejiang and Shanghai were 285.542 billion yuan, 218.042 billion yuan and 182.220 billion yuan respectively. In 2019, except for the slight increase (0.95%) in the total assets of Zhejiang NEEQ listed companies, the total assets in other regions have all decreased to varying degrees, though relatively low in Guangdong, Jiangsu and Shanghai (See Fig. 6.2).

Judging from the NEEQ listed companies that have operated for two consecutive years from 2018 to 2019, except for the slight decrease (4.77%) in the total assets of the companies in Zhongguancun, the total assets in other regions have all increased to varying degrees. Among them, the total assets of continuously operating companies in Zhejiang and Shanghai have increased by more than 10%.

6.1.3 Financing

1. Stock Issue

According to the data disclosed by National Equities Exchange and Quotations, the total amount of equity financing of NEEQ listed companies across the country was

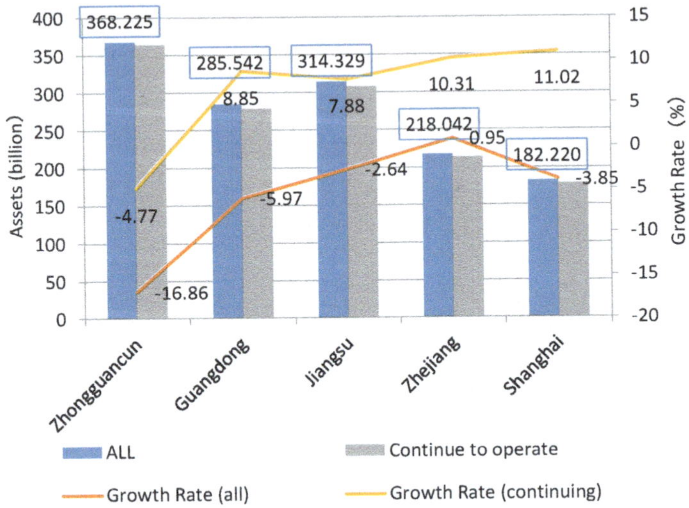

Fig. 6.2 The total assets of the NEEQ listed companies in each region in 2019. *Source* Wind, compiled by ZLCA

26.463 billion yuan in 2019, a significant decrease of 128% compared with that of 2018 (60.443 billion yuan). In line with the national trend, the financing situation in the five major regions was not optimistic in 2019. Among them, Zhongguancun stock was issued 79 times, with a financing amount of 4.099 billion yuan (a 50.62% year-on-year decline). Guangdong stock was issued 91 times, raising 3.687 billion yuan (a 51.72% year-on-year decline). The amount of equity financing of Jiangsu, Zhejiang and Shanghai in 2019 were 5.368 billion yuan, 1.312 billion yuan and 1.366 billion yuan respectively (see Table 6.2).

In terms of the way of stock issuance, Zhongguancun NEEQ listed companies conducted 67 private placements in 2019, with 700 million shares accounting for 3.740 billion yuan, down 20.66% year on year. Guangdong NEEQ listed companies conducted 85 private placements in 2019, with 396 million shares accounting for

Table 6.2 Stock Issue of NEEQ listed companies in major regions in 2019

Region	2019		2018		Growth Rate(%)	
	Amount (billion yuan)	Times	Amount (billion yuan)	Times	Amount	Times
Zhongguancun	4.099	79	8.301	186	−50.62	−57.53
Guangdong	3.687	91	7.636	253	−51.72	−64.03
Jiangsu	5.368	92	7.624	171	−29.59	−46.20
Zhejiang	1.322	47	3.828	105	−65.46	−55.24
Shanghai	1.366	53	4.706	110	−70.97	−51.82

Source NEEQ, compiled by ZLCA

Table 6.3 Comparison of private placements by NEEQ listed companies in major regions in 2019

Regions	Number of additional issues		Number of additional issues		Financing amount	
	Times (times)	Growth rate (%)	Quantity (million shares)	Growth rate (%)	Amount (billion yuan)	Growth rate (%)
Zhongguancun	67	−53.47	700	−16.32	3.740	−20.66
Guangdong	85	−55.73	396	−65.19	1.907	−63.41
Jiangsu	83	−44.30	2851	97.84	5.898	−12.59
Zhejiang	42	−53.33	315	−53.00	1.394	−55.82
Shanghai	46	−45.24	231	−66.29	1.312	−58.89

Source Wind, compiled by ZLCA

1.907 billion yuan. Jiangsu conducted 83 private placements, with 2.851 billion shares accounting for 5.898 billion yuan. Both the number of issuance and the amount of financing in the region outperformed that in other regions. Zhejiang and Shanghai conducted 42 and 46 private placements respectively, with financing amounts of 1.394 billion yuan and 1.312 billion yuan (see Table 6.3).

2. Bond Issue

In 2019, a total of 6 NEEQ listed companies nationwide issued 9 bonds, with a total financing amount of 5.510 billion yuan. Among them, three companies in Zhongguancun, including China Kangfu International Leasing Co., Ltd., China National Investment and Guaranty Corporation and Beijing Pan-China Sports Industry Co.,Ltd, issued 6 bonds totaling 5.34 billion yuan, with an average coupon rate of 5.91%. There were 2 companies in Zhejiang that issued bonds with a total scale of 150 million yuan, with an average coupon rate of 6.90%. Only one company in Guangdong (Guangdong Flying Enterprise Internet Technology Co., Ltd) has issued bonds, with the issuance scale of 20 million yuan and coupon rate of 6.50%. In 2019, the number of NEEQ listed companies and the actual amount of bond financing in Zhongguancun respectively accounted for 66.67% and 96.91% of the country's total, much higher than that of other regions. When the bond financing in the overall NEEQ market was not active, Zhongguancun NEEQ listed companies outperformed those in other regions.

6.1.4 Tax Payment

6.1.4.1 Enterprise Income Tax

In 2019, the enterpriseincome tax payment of Zhongguancun's NEEQ listed companies totaled 2.0 billion yuan, basically the same as that of Guangdong, Jiangsu and

Zhejiang. The total income tax payment of Shanghai NEEQ listed companies was relatively low, at 0.91 billion yuan. At the same time, compared with 2018, the average income tax paid by NEEQ listed companies in Zhongguancun and Shanghai has also decreased significantly to 1.8501 million yuan and 1.3602 million yuan respectively (2.1993 million yuan and 1.8901 million yuan in 2018). Judging from the growth rate, Shanghai and Zhongguancun experienced the largest decrease (36.98% and 25.71% respectively) in income tax payment. The income tax payment of Jiangsu NEEQ listed companies has also fallen by over 10%, while Guangdong and Zhejiang remained basically the same as last year (See Fig. 6.3).

Judging from the income tax paid by NEEQ listed companies that continued to operate from 2018 to 2019, the largest income tax reductions were in zhongguancun and Shanghai. In 2019, the total profit of the continuously operating NEEQ listed companies in Zhongguancun has increased by 21.96%, while the income tax paid has decreased by 6.06%; the total profit of the continuously operating companies in Shanghai has increased by 10.94%, while the income tax paid has fallen by 15.65%. The income taxes paid by the continuously operating NEEQ listed companies in Guangdong, Jiangsu and Zhejiang have all increased to a certain extent, especially in Guangdong, where the growth rate even reached 38.89% (while the total profit increased by 29.97% year-on-year). In terms of the proportion of income tax to total profits, the proportion of tax paid by the continuously operating companies in Jiangsu has been the lowest over the past two years, at around 18%; the proportion of income tax in Zhongguancun and Shanghai was 23.18% and 22.60% respectively, both representing a significant decrease compared with 2018 (see Table 6.4).

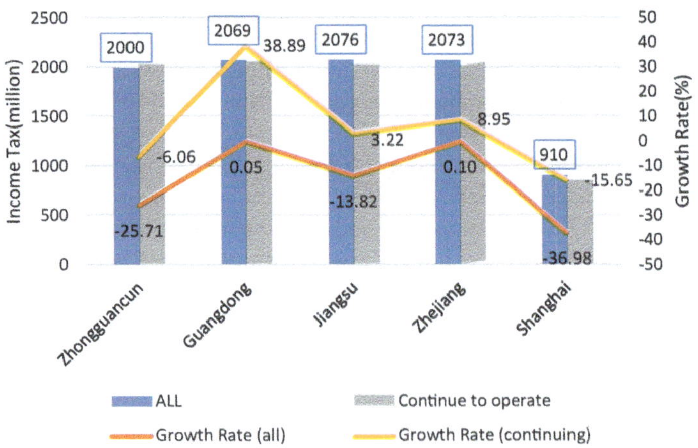

Fig. 6.3 Income tax of NEEQ listed companies in major regions in 2019. *Source* Wind, compiled by ZLCA

Table 6.4 Income tax of continuously operating NEEQ listed companies in major regions in 2019

Region	The total profit		Enterprise income tax		Proportion of income tax to total profit (%)	
	Amount (billion yuan)	Growth rate (%)	Amount (billion yuan)	Growth rate (%)	2018	2019
Zhongguancun	8.753	21.96	2.029	−6.06	30.10	23.18
Guangdong	8.712	29.97	2.050	38.89	22.02	23.53
Jiangsu	11.026	−4.11	2.020	3.22	17.02	18.32
Zhejiang	10.695	14.24	2.008	8.95	19.69	18.78
Shanghai	3.743	10.94	0.846	−15.65	29.73	22.60

Source Wind, compiled by ZLCA

6.1.4.2 Actual Tax Burden

The actual tax burden of Zhongguancun NEEQ listed companies in 2019 was 7.137 billion yuan, a decrease of 24.23% compared with 9.419 billion yuan in 2018. It was the largest drop. The actual tax burdens in Guangdong, Jiangsu, Zhejiang and Shanghai were respectively 5.711 billion yuan, 6.140 billion yuan, 4.447 billion yuan and 4.094 billion yuan, all with a year-on-year decrease of more than 10% (even greater than that of 2018) (See Fig. 6.4). In terms of the average actual tax burden of enterprises, except for Shanghai, where the average tax burden in 2019 was basically the same as the previous year, the average actual tax burden in other major regions all experienced a relatively significant decline. Among them, the average actual tax burden of Zhongguancun NEEQ listed companies decreased from 7.6953 million

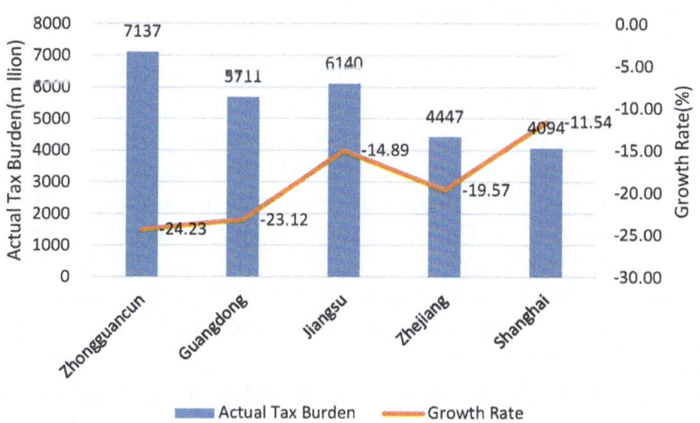

Fig. 6.4 Actual tax burden of NEEQ listed companies in major regions in 2019. *Source* Wind, compiled by ZLCA

Table 6.5 Proportion of actual tax burden to revenue of NEEQ listed companies in major regions in 2019

Region	Revenue		Proportion of actual tax burden to revenue (%)		
	Amount (billion yuan)	Growth rate (%)	2018	2019	Growth rate
Zhongguancun	206.6826	−13.64	3.94	3.45	−12.36
Guangdong	209.4011	−12.85	3.09	2.73	−11.70
Jiangsu	181.6375	−9.20	3.61	3.38	−6.36
Zhejiang	219.3304	3.52	2.61	2.03	−22.32
Shanghai	265.2538	3.77	1.81	1.54	−14.73

Source Wind, compiled by ZLCA

yuan to 6.6022 million yuan, representing the largest drop of 14.20%; Guangdong and Zhejiang also fell by more than 10%.

Judging from the proportion of the actual tax burden to the revenue, the proportion of NEEQ listed companies in major regions has all shown a downward trend, but the proportion of Zhongguancun NEEQ listed companies is still the highest of all. In 2019, the proportion of the actual tax burden to the revenue in Zhongguancun, Guangdong, Jiangsu, Zhejiang and Shanghai was 3.45%, 2.73%, 3.38%, 2.03% and 1.54% respectively (see Table 6.5).

6.1.5 *Staff Composition*

6.1.5.1 Total Number of Employees

The number of employees in Zhongguancun NEEQ listed companies was 171,000 in 2019, which was only higher than the number of employees in NEEQ listed companies in Shanghai (134,400). The number of employees in Guangdong, Jiangsu and Zhejiang in 2019 was respectively 295,500, 210,100 and 179,500. In terms of growth rate, the number of employees in Zhongguancun has reduced by 59,800 in 2019, a decrease of 25.91%. The decrease in the number of employees in Jiangsu, Zhejiang and Shanghai was kept between 10 and 20%, basically the same rate as in 2018. The number of employees in Guangdong in 2019 still maintained a downward trend, but the decline has slowed down, from 30.98% in 2018 to 17.32% in 2019 (See Fig. 6.5).

In terms of the number of employees in NEEQ listed companies that have continued operations for two consecutive years from 2018 to 2019, the number of employees in Zhongguancun, Guangdong and Jiangsu remained basically the same as last year. The number of employees in the continuously operating companies in Shanghai has slightly decreased (3.29%), while the number of employees in Zhejiang

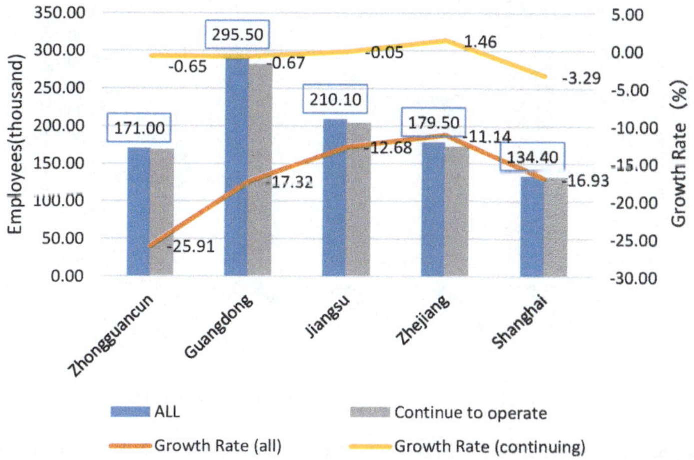

Fig. 6.5 Total number of employees in NEEQ listed companies in major regions in 2019. *Source* Wind, compiled by ZLCA

has increased slightly (1.46%). It can be seen that the decline in the total number of employees in NEEQ listed companies in the above five regions is mainly caused by the reduction in the number of companies rather than the layoffs in companies.

6.1.5.2 Employees with a Higher Degree Than a Specialist

In 2019, the number of employees with a higher degree than a specialist was 80,000 (69,722 undergraduates, 9,662 masters, and 614 doctors) in Zhongguancun NEEQ listed companies, which accounted for 46.78% of the total with the average number being 74. Guangdong ranked second with a total number of 69,400 (63,677 undergraduates, 5,348 masters and 375 doctors), which accounted for 23.49% of the total with the average number being 58. The total number of staff with higher degree in Jiangsu, Zhejiang and Shanghai was respectively 46,300, 33,600 and 51,000 (see Fig. 6.6). The data showed that the overall education background of the employees in Zhongguancun NEEQ listed companies was better than that in other regions. Both the number of undergraduates, masters and doctors calculated alone and the total number were significantly higher than those in other regions. It proved that Zhongguancun NEEQ listed companies, whose employees have higher quality and are more competitive, can attract more highly-educated and high-quality talents.

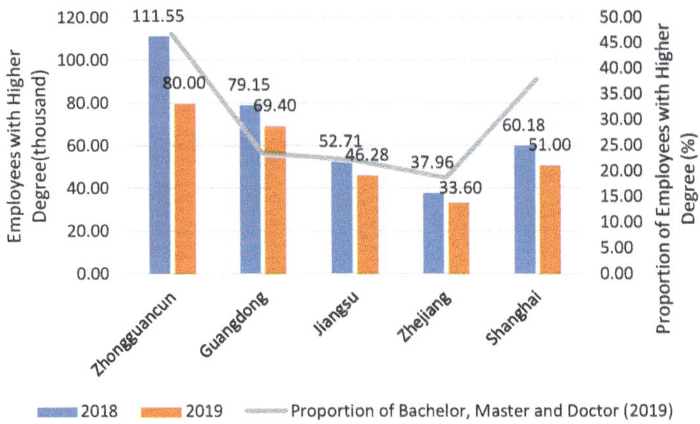

Fig. 6.6 Number and proportion of employees with higher degree in NEEQ listed companies in major regions in 2019. *Source* Wind, compiled by ZLCA

6.1.6 Summary

Through the analysis of the market cap, assets, financing, tax payment and staff composition in major regions of the country, it can be seen that in 2019 when the overall macroeconomy suffered continuous downturn, the number of NEEQ delisted companies nationwide has increased, and all the above indicators in the major regions have decreased. But the decrease in those indicators is mainly caused by the decline in the number of companies rather than the drop in the performance of continuously operating companies.

1. While under huge downward pressure, the total market cap and total assets of Zhongguancun NEEQ listed companies were still ahead of those in other regions. In 2019, although the total market cap and total assets in major regions all showed a downward trend and the decline in Zhongguancun was larger than those in other regions, Zhongguancun still outperformed other regions. It shows that Zhongguancun NEEQ listed companies have a solid foundation and investors have great confidence in them.
2. In 2019, the financing amount of NEEQ listed companies across the country has all decreased significantly compared with 2018. NEEQ listed companies' financing capacity has weakened, but the overall financing capacity of Zhongguancun were still ahead, especially in bond issuance, where Zhongguancun has an absolute advantage.
3. In 2019, the proportion of average actual tax burden to revenue in major regions have all decreased, but to varying degrees. Zhongguancun has imposed lower taxes on its NEEQ listed companies that have operated for two consecutive years from 2018 to 2019. Both the corporate income tax and the proportion of income tax to total profits in the region have decreased significantly, but the proportion of income tax was still relatively high (approximately 5 percentage points higher

than Jiangsu, where the proportion was the lowest). Therefore, the taxes imposed on Zhongguancun NEEQ listed companies should be further reduced.

4. In 2019, the total number of employees, the average number of employees and the average number of employees with higher degree in major regions have all decreased to varying degrees. The total number and the proportion of employees with higher degree, and the average number of employees with higher degree in Zhongguancun were all higher than those in other regions, which was indicative of the high quality of the staff.
5. In 2019, the decline in market capitalization, assets and number of employees in major regions across the country was mainly due to the decrease in the number of companies. The NEEQ listed companies that continued their operation from 2018 to 2019 in major regions have performed relatively well. The growth momentum was strong, and the various indicators remained strong as well, basically the same as in 2018 or with slight increase.

6.2 Analysis of Profitability of NEEQ Listed Companies in Different Regions

6.2.1 Revenue

In 2019, the total revenue of Zhongguancun NEEQ listed companies was 206.683 billion yuan, higher than the 181.638 billion yuan of Jiangsu, but lower than the total revenues of Shanghai, Zhejiang and Guangdong. In terms of growth, except for Zhejiang and Shanghai, where the total revenues increased, the total revenues in other regions in 2019 have all declined to varying degrees. In terms of the average revenue in 2019, Shanghai NEEQ listed companies were ahead of those companies in other regions, amounting to 396 million yuan. The average revenue in Guangdong was the lowest, at 175 million yuan. And Zhongguancun was in a middle position in terms of the average revenue of its NEEQ listed companies, amounting to 191 million yuan.

Judging from the total revenue of NEEQ listed companies that have operated for two consecutive years from 2018 to 2019, the NEEQ listed companies in the above five regions have all witnessed increases in 2019. Among them, the increase rate in Zhongguancun and Guangdong was almost the same, both at about 6%. The total revenue of continuously operating companies in Jiangsu and Zhejiang has both increased by about 9%, while the increase rate in Shanghai was the largest, reaching 18.07% (See Fig. 6.7).

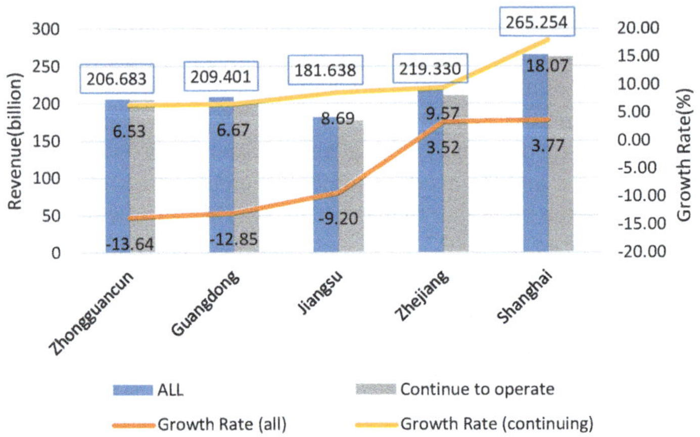

Fig. 6.7 Total revenue of NEEQ listed companies and continuously operating NEEQ listed companies in major regions in 2019. *Source* Wind, compiled by ZLCA

6.2.2 *Profits*

6.2.2.1 Gross Profit

The total gross profit of Zhongguancun NEEQ listed companies was 46.631 billion yuan in 2019, a year-on-year decrease of 18.80%. The gross profit of Guangdong, Jiangsu, Zhejiang and Shanghai was respectively 50.572 billion yuan, 40.312 billion yuan, 35.175 billion yuan and 28.255 billion yuan. Similar to Zhongguancun, the gross profit in other major regions in 2019 all showed a downward trend. The smallest decline was in Zhejiang, which was 2.32%. And the growth rates of Guangdong, Jiangsu and Shanghai were respectively −11.77%, −8.50% and −12.25% (see Fig. 6.8). The data indicates that the gross profit of Zhongguancun NEEQ listed companies in 2019 is slightly lower than that of Guangdong and higher than that of other regions.

The gross profit and gross margin of Zhongguancun NEEQ listed companies in 2019 were respectively 43 million yuan and 22.56%, which meant that Zhongguancun had strong profitability, giving it the edge over other regions. On the whole, in 2019, the average gross profits of NEEQ listed companies in majors regions of the country differed a little, all ranging between 40 million yuan and 50 million yuan. But the gross margin of Shanghai and Zhejiang is obviously lower than those in other regions. (See Table 6.6).

6.2.2.2 Net Profit

In 2019, the total net profit of Zhongguancun NEEQ listed companies was 8.137 billion yuan, lower than the 9.128 billion yuan of Jiangsu and 8.998 billion yuan

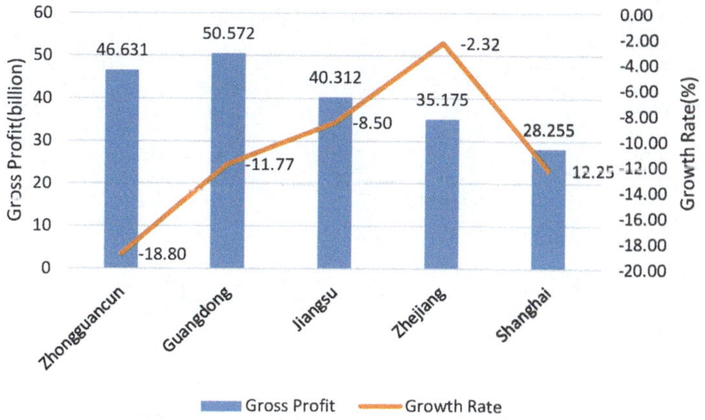

Fig. 6.8 Gross profit of NEEQ listed companies in major regions in 2019. *Source* Wind, compiled by ZLCA

Table 6.6 Comparison of profitability of NEEQ listed companies in major regions in 2019

Region	Total revenue (billion yuan)	Average revenue (million yuan)	Revenue per capita (thousand yuan)	Gross profit (billion yuan)	Average gross profit v	Gross margin (%)	Net profit (billion yuan)	Net profit per capita (thousand yuan)
Zhongguancun	206.683	191	1208.8	46.631	43	22.56	8.137	47.6
Guangdong	209.401	175	508.9	50.572	42	24.15	6.279	21.2
Jiangsu	181.638	185	864.7	40.312	41	22.19	9.128	43.5
Zhejiang	219.330	300	1221.8	35.175	48	16.04	8.998	50.1
Shanghai	265.254	396	1974.1	28.255	42	10.65	3.017	22.5

Source Wind, compiled by ZLCA

of Zhejiang. The net profits of Guangdong and Shanghai regions were respectively 6.279 billion yuan and 3.017 billion yuan. Except for Zhejiang, where the total net profit of the NEEQ listed companies has increased a little (6.64%), the net profits in other regions have all dropped significantly in 2019. Among them, Shanghai had the largest decline, amounting to 31.88%. The net profits of NEEQ listed companies in Guangdong and Zhejiang also fell by more than 20%. Zhongguancun had a relatively small decline, amounting to 11.82% (See Fig. 6.9). In terms of average net profit and net profit per capita, Zhongguancun NEEQ listed companies had certain advantages over companies in other regions, with the average net profit being 7.5273 million yuan, higher than that of Guangdong and Shanghai. The net profit per capita in Zhongguancun was 47,600 yuan, second only to the 50,100 yuan in Zhejiang.

Judging from the net profit of NEEQ listed companies that have continued operations for two consecutive years from 2018 to 2019, the net profits of the continuously operating NEEQ listed companies in Zhongguancun and Guangdong were higher

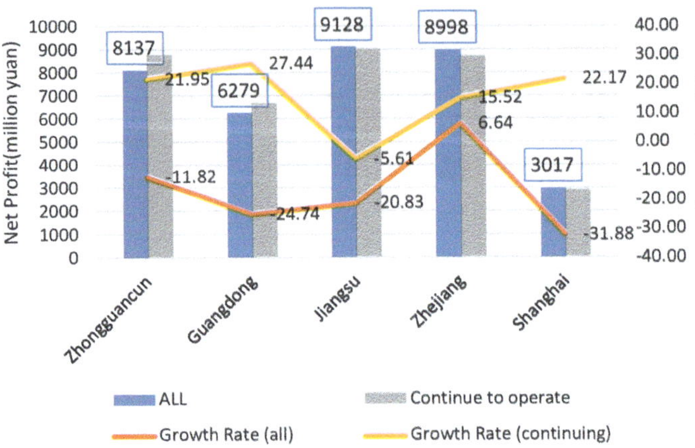

Fig. 6.9 Net profit of NEEQ listed companies in major regions in 2019. *Source* Wind, compiled by ZLCA

than the net profits of all NEEQ listed companies in the two regions. The net profit of Zhongguancun was 8.753 billion yuan, representing an increase of 21.95%. The net profit of Guangdong was 6.662 billion yuan, representing the largest increase of 27.44%. The net profits of the continuously operating NEEQ listed companies in Zhejiang and Shanghai have all increased by more than 15%, while the net profit of Jiangsu has declined slightly, with a decrease of 5.61%.

6.2.3 *Summary*

From the perspective of corporate profitability, the NEEQ listed companies in Shanghai and Zhejiang have delivered better performance. Zhejiang has achieved growth both in revenue and net profit, and Shanghai has maintained the growth in revenue. Considering the NEEQ listed companies that continued to operate from 2018 to 2019, Zhongguancun, Guangdong, Zhejiang and Shanghai have all achieved growth both in revenue and net profit. The overall decline in the revenue and the net profit of all the Zhonguancun NEEQ listed companies was mainly caused by the decrease in the number of enterprises or the loss of new enterprises. Though already in a leading position, the profitability of Zhongguancun NEEQ listed companies should be further improved.

Table 6.7 Preparations for select in major regions (unit Home)

Region	Number of guidance filing	Number of guidance acceptance	Number of acceptance	Number of approval in review	Number of approval by china securities regulatory commission
Zhongguancun	25	12	11	6	4
Guangdong	20	6	4	2	2
Jiangsu	19	6	6	5	4
Zhejiang	14	5	5	2	2
Shanghai	13	5	5	2	1
Nationwide	171	75	68	32	24

Source Local Financial Supervision and Administration, compiled by ZLCA

6.3 Analysis of Innovation in Different Regions

6.3.1 Analysis of Preparations for Select in Each Region

As of July 1, 2020, there were a total of 171 enterprises nationwide that have applied for guidance filing (guidance period and acceptance), among which 25 were Zhongguancun enterprises, accounting for 14.62% (ranking first). 20 enterprises were in Guangdong. Jiangsu, Zhejiang and Shanghai boasted between 10 and 20 enterprises that have applied for guidance filing. In terms of the progress, a total of 96 enterprises across the country were in guidance period, and 75 enterprises have completed guidance acceptance, among which 12 were in Zhongguancun (ranking first), 6 in Guangdong and Jiangsu respectively, and 5 in Zhejiang and Shanghai respectively. In terms of application and acceptance of Select, National Equities Exchange and Quotations has accepted the applications from 68 enterprises, with 11 in Zhongguancun (ranking first). In terms of review of Select, as of July 1, 2020, 32 enterprises have been approved in the review of application for the Select, among which 6 were in Zhongguancun (ranking first). As for the number of approvals by China Securities Regulatory Commission, as of July 1, 2020, both Zhongguancun and Jiangsu ranked first, each with four enterprises (Table 6.7).

6.3.2 Analysis of Core Indicators of Enterprises on Innovation in Different Regions

In 2019, there were 637 innovation-level enterprises across the country. Among them, there were 90 in Zhongguancun, accounting for 14.13% of the total. There were 88 in Guangdong, accounting for 13.81%; and there were respectively 63, 51 and 56 innovation-level enterprises in Jiangsu, Zhejiang and Shanghai (see Fig. 6.10).

In terms of the proportion of innovation-level enterprises to all the NEEQ listed companies in the region, the proportion in Zhongguancun was 8.33%, slightly lower than the 8.37% in Shanghai. The proportion in Guangdong, Jiangsu and Zhejiang was respectively 7.33%, 6.40% and 6.99% (See Table 6.8). Data shows that the number of innovation-level enterprises in Zhongguancun and the proportion of it to the innovation-level enterprises across the country were both higher than those of other regions, which was indicative of the high quality of Zhongguancun NEEQ listed companies.

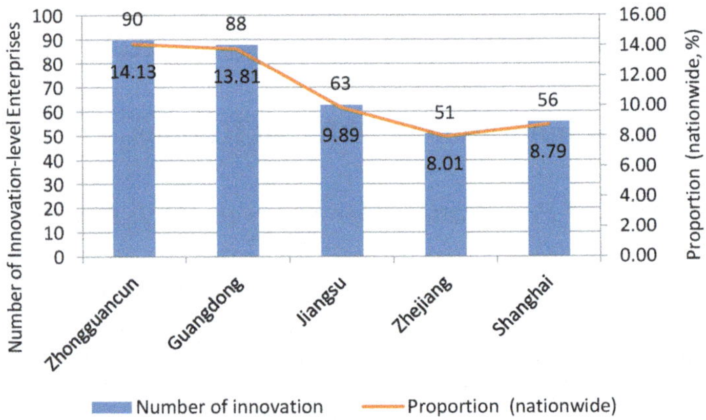

Fig. 6.10 Number and proportion of innovation-level enterprises in major regions in 2019. *Source* Wind, compiled by ZLCA

Table 6.8 Comparative analysis of innovation-level enterprises in major regions in 2019

Region	Number of innovation	Proportion (innovation ofnatiowide, %)	Proportion (corresponding region, %)	Market cap (billion yuan)	Total assets (billion yuan)	Total revenue (billion yuan)	Net profit (billion yuan)
Zhongguancun	90	14.13	8.33	73.017	71.700	47.772	1.929
Guangdong	88	13.81	7.33	66.465	62.435	45.830	2.953
Jiangsu	63	9.89	6.40	39.235	41.066	21.239	1.597
Zhejiang	51	8.01	6.99	50.497	67.771	48.850	1.998
Shanghai	56	8.79	8.37	44.517	41.107	144.019	2.088

Source Wind, compiled by ZLCA

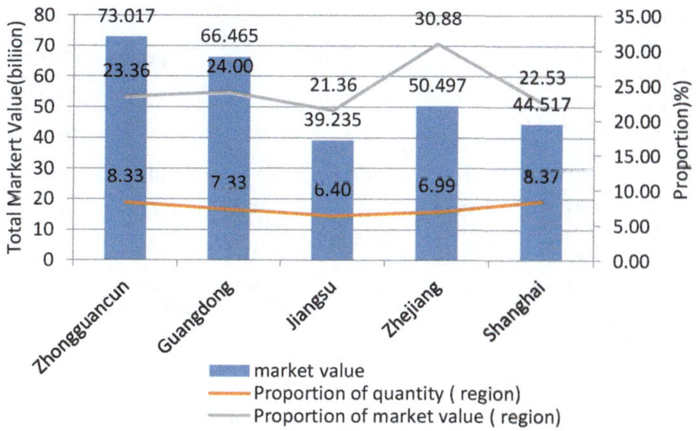

Fig. 6.11 Market cap and the proportion of innovation-level companies in major regions in 2019. *Source* Wind, compiled by ZLCA

6.3.2.1 Market Capitalization

In 2019, the market cap of innovation-level companies of Zhongguancun was much higher than that of other regions, amounting to 73.017 billion yuan, followed by 66.465 billion yuan of Guangdong. The market cap of innovation-level NEEQ listed companies in Jiangsu, Zhejiang and Shanghai was respectively 39.235 billion yuan, 50.497 billion yuan and 44.517 billion yuan. At the same time, the innovation-level NEEQ listed companies in Zhongguancun, which accounted for less than 10% of all the NEEQ listed companies in the region, have contributed to more than 20% of the market cap, and the proportion of the market cap of innovation-level companies in Zhejiang even reached 30.88% (See Fig. 6.11). The absolute advantages in both the number of innovation-level enterprises and the market cap also reflect the advantage in quality that Zhongguancun NEEQ listed companies have over other regions.

6.3.2.2 Assets and Revenue

The total assets of Zhongguancun's innovation-level enterprises in 2019 were 71.700 billion yuan, which was ahead of other regions. The average total assets was 797 million yuan, lower than that of Zhejiang and higher than that of other regions. The total assets of innovation-level enterprises in Guangdong, Jiangsu, Zhejiang and Shanghai in 2019 were respectively 62.435 billion yuan, 41.06 billion yuan, 67.771 billion yuan and 41.107 billion yuan. In terms of revenue, the total revenue of Zhongguancun's innovation-level companies was 47.772 billion yuan, lower than the 144.019 billion yuan of Shanghai and 48.850 billion yuan of Zhejiang. The average revenue of Zhongguancun's innovation-level companies was 531 million yuan, lower than the 2,572 million yuan of Shanghai and 958 million yuan of Zhejiang, which

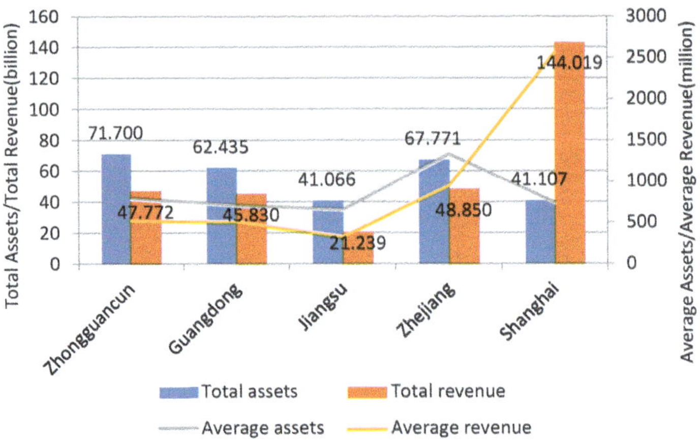

Fig. 6.12 Assets and revenue of innovation-level companies in various regions in 2019. *Source* Wind, compiled by ZLCA

indicates the relatively small size of the innovation-level companies in Zhongguancun (see Fig. 6.12).

6.3.2.3 Net Profit

In 2019, the net profit of Zhongguancun's innovation-level companies was 1.929 billion yuan, which contributed to 23.71% of the total profit in the region, showing strong profitability. The total net profits of the innovation-level NEEQ listed companies in Guangdong, Jiangsu, Zhejiang and Shanghai in 2019 were respectively 2.953 billion yuan, 1.597 billion yuan, 1.998 billion yuan and 2.088 billion yuan. Among them, the profitability of Shanghai's innovation-level companies was remarkably stronger than that of the companies on the Base. The innovation-level companies in Shanghai, which accounted for only 8% of the total, contributed to about 70% of the net profit in the region. In terms of profitability, in 2019, 60 innovation-level companies in Zhongguancun have achieved profitability, with the proportion being 66.67% (the total number of innovation-level companies), which was relatively low compared with other regions. The total net profit of those 60 innovation-level companies was 3.042 billion yuan, with the average net profit being 51 million yuan. The proportion (to the total number of innovation-level companies in the corresponding region) of innovation-level NEEQ listed companies in Jiangsu and Shanghai that have achieved profitability in 2019 was relatively high, 87.30% and 87.50% respectively (See Table 6.9). Zhongguancun's innovation-level NEEQ listed companies have certain advantages in net profit, but those companies vary in quality, a large proportion of which being unprofitable companies and some being companies with heavy losses.

Table 6.9 Comparison of profitability of innovation-level enterprises in various regions in 2019

Region	Net profit (billion yuan)	Number of profitable companies	Proportion of profitable companies (%)	Net profit of profitable companies (billion yuan)	Amount of losses (million yuan)	Average losses (million yuan)
Zhongguancun	1.929	60	66.67	3.042	1113	37
Guangdong	2.953	70	79.55	3.440	487	27
Jiangsu	1.597	55	87.30	1.929	332	42
Zhejiang	1.998	39	76.47	2.592	594	50
Shanghai	2.088	49	87.50	2.269	181	26

Source Wind, compiled by ZLCA

6.3.3 Analysis of Innovation Ability in Various Regions

6.3.3.1 Innovation Resources

In 2019, Zhongguancun NEEQ listed companies had a total of 10,276 masters and doctors, accounting for 6.01% of all employees, a slight increase from 5.94% in 2018. Zhonguancun was followed by Guangdong, whose number of masters and doctors was 5,723 and the proportion was 1.94%. The number of masters and doctors in Jiangsu, Zhejiang and Shanghai was respectively 4,822, 3,102 and 5,632, respectively accounting for 2.30, 1.73 and 4.19% (see Fig. 6.13). Both the number and the proportion of masters and doctors in Zhongguancun are much higher than those of other regions, which enables Zhonguancun to attract more high-quality talents and have greater potential for innovation.

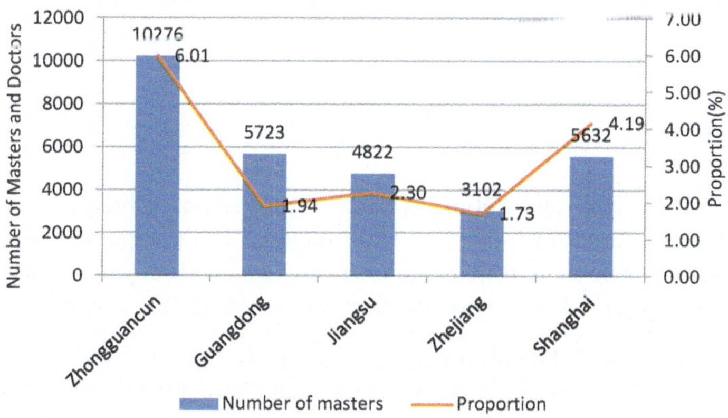

Fig. 6.13 Analysis of number and proportion of masters and doctors in major regions in 2019. *Source* Wind, compiled by ZLCA

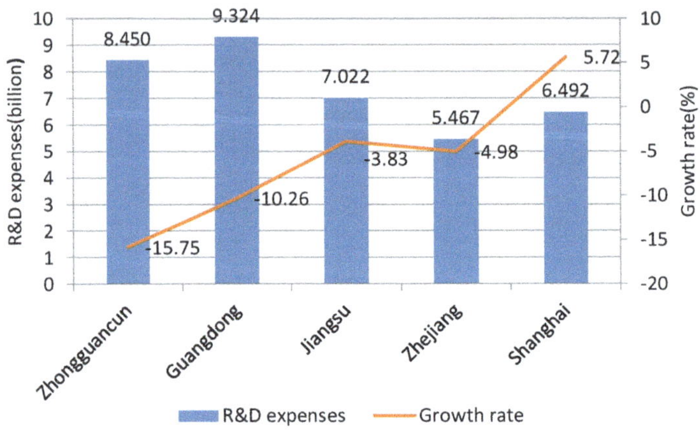

Fig. 6.14 R&D expenses in major regions in 2019. *Source* Wind, compiled by ZLCA

6.3.3.2 Innovation Drive

In 2019, the research and development expenses of Zhongguancun NEEQ listed companies totaled 8.45 billion yuan, and the average R&D expenses was 7.8167 million yuan, with the average R&D intensity reaching 4.70%. The total R&D expenditure in Guangdong was 9.324 billion yuan, which was ahead of other major regions, with the average R&D expenses being 7.7697 million yuan and the average intensity being 5.73%. The total R&D expenses in Jiangsu, Zhejiang and Shanghai were respectively 7.022 billion yuan, 5.467 billion yuan and 6.492 billion yuan, with the average expenses being 7.1362 million yuan, 7.4894 million yuan and 9.704 million yuan (see Fig. 6.14). The total R&D expenditure of Zhongguancun was slightly lower than that of Guangdong, but higher than other regions, and the average R&D expenditure and R&D intensity were both at the intermediate level.

6.3.4 Summary

Judging from the innovation ability of the major regions, Zhongguancun NEEQ listed companies outperform those in other regions. The competitive advantage of Zhongguancun can be seen from the number ofSelect and Innovation, the number of employees with higher degree, innovation resources of R&D and innovation drivers.

1. As of July 1, 2020, the number of applications for guidance filing, of guidance acceptance, of approvals in listing review and of approvals by China Securities Regulatory Commission in Zhongguancun were all higher than those of other regions. Nutrichem, a Zhongguancun NEEQ listed company engaged in the research and development, production and sales of high-quality pesticide

intermediates, primary agricultural chemicals and pesticide preparations, even became the first NEEQ listed company that has entered into the Select.

2. The number, the market cap and the total assets of Zhongguancun's innovation-level NEEQ listed companies were all higher than those of other regions. However, due to differences in corporate characteristics, the total revenue of Shanghai's innovation-level NEEQ listed companies was significantly higher than those of other regions, which was almost three times as high as the total revenue of Zhongguancun. The size of Shanghai enterprises was larger. In terms of net profit, Guangdong's innovation-level NEEQ listed companies outperformed those of other regions.
3. Zhongguancun NEEQ listed companies were ahead of those of other regions in terms of innovation resources (the number and proportion of employees with a higher degree than a specialist), innovation driver (R&D investment), etc., but they were not so competitive in some other aspects. For instance, the overall R&D expenditure of Zhongguancun NEEQ listed companies was slightly lower than that of Guangdong, and the innovation ability of Zhongguancun should be further improved.

Chapter 7
The Analysis of Potentiality of the Zhongguancun NEEQ Companies to 'Select'

Abstract 2019 has been a significant year for NEEQ. The reform of NEEQ with aspect to the relevant policies and regulations were continuously implemented since China Securties Regulatory Commission has announced the deepen reform of NEEQ on 25th October 2019. The deepen reform of NEEQ has already got into a new stage. The listed NEEQ companies are meeting the significant opportunities. In this report, the situations of the Zhongguancun companies which meets the criterions to be listed in 'Select' layer is analyzed according to the financial ratios in NEEQ.

Keyword Zhongguancun NEEQ Select

7.1 The Potentiality Analysis of Zhongguancun 'Select' Layer Companies

7.1.1 The Significance and Background of 'SELECT' Layer

NEEQ is the important platform to provide the financing, trading and pricing services to the small and medium-sized companies in the Multi-level capital market. NEEQ develops in a different method with the Shanghai and Shenzhen stock exchanges. A diversified and complimentary system of capital market is constituted by them together. The comprehensive deepen reform has brought the historic opportunities to the NEEQ market which helps the different layers of market can be developed orderly. The functionality of the NEEQ market can be further improved and the position of the NEEQ market to serve the innovative small-medium sized companies is more precise. For companies, they have a clearer expectation and direction towards the future development in the capital market.

The reform of NEEQ concentrates on the Layered management method, Management measures for the appropriateness of investors, Public offering rules, Directional issuing rules, Stock trading rules, Board transfer rules, information disclosure rules, supervision and management measures, listed companies' governance rules so on. The most market attention is caught by the establishment of 'Select' layer as a layered

Zhongguancun Listed Companies Association, *The Growth Report of Zhongguancun NEEQ Listed Companies (2020)*, Current Chinese Economic Report Series,
https://doi.org/10.1007/978-981-33-6819-4_7

management method. This has formed the more comprehensive market structure, such as 'Base', 'Innovation' and 'Select' layer. Furthermore, the mechanism of board transformation is designed to construct a dynamic multi-layered capital market.

7.1.2 The Analysis of the Pre-Select Zhongguancun NEEQ Listed Companies

The NEEQ market was established in the Zhongguancun Industrial park. The number of NEEQ listed companies in Zhongguancun industrial park is more than the same-type industrial park with the years of NEEQ development and expansion. Until the date of 31st December 2019, there are the number of 1190 Zhongguancun companied listed in NEEQ, accounts for 13.29% of the total 8953 NEEQ listed companies in China. Zhongguancun industrial park has become the necessary component of the NEEQ market. The industrial categories of Zhongguancun NEEQ listed companies are diversified. They have covered various industries, such as Information transmission, Software and IT services, Scientific research, technology service, Manufacturing, Leasing and Business services, Culture, Sports and Entertainment and Education. The level of innovation ability for Zhongguancun companies is higher than the national average level.

The reform of NEEQ has started since 25th October 2019, Zhongguancun Listed Company Association NEEQ branch actively carried out the work in order to support the reform. At the beginning of the reform, the NEEQ branch reports to the Zhongguancun Management Committee and the NEEQ company about the companies 'attitude, view and propositions toward the NEEQ reform by organizing the reform workshops, key enterprise communication and enterprise site-investigation. In the mid-term of the reform, the NEEQ Branch sorted out the potential Zhongguancun companies which meets the requirements to be listed in 'Select' layer according to the different financial indexes. In addition, the NEEQ branch organized the relevant workshops, assist the companies in applying for themselves to be listed in 'Select' layer, online or offline site-investigated the key companies about the progress. At the later stage, the NEEQ Branch, Zhongguancun management committee, Finance offices and National NEEQ company took extra efforts to site-investigate the guidance-recorded companies, tried to find out the problems the companies encountered when they are in the process of application to be listed in 'Select' layer and helped them to solve them.

In a common endeavor of the Zhongguancun management community, Finance Offices and the NEEQ company, the Zhongguancun NEEQ listed companies has win the first price in the reform of 'Select' layer. Until the 30th June 2020, there are the number of 169 guidance-recorded companies, including the 25 Zhongguancun companies (ranks 1st), accounts for 15.24%; In the perspective of the guidance process, there are 94 companies in the stage of guidance, and 75 companies have completed the guidance acceptance, including the 12 Zhongguancun companies (ranks 1st); In the perspective of the application and acceptance, the NEEQ system has accepted 63 companies' application, including the 10 Zhongguancun companies (ranks 1st); In the perspective of List auditing, there are 32 companies have passed the list auditing by 1st July 2020 including the 6 Zhongguancun companies (ranks 1st). The 25 Zhongguancun NEEQ listed companies which applied for the guidance recording are shown following (see Table 7.1).

Table 7.1 The application status of Zhongguancn NEEQ listed companies to 'Select' layer

Serial number	Stock code	Company name	Listing date	Category	Host broker	Preparation progress	Listing committee audit results
1	833,819.OC	Nutrichem Company Limited	2015/10/20	Manufacturing	Southwest securities	Application accepted	Passed
2	832,317.OC	Skycam Defense Technology Co., Ltd.	2015/4/15	Scientific research and technology service	CITIC securities	Application accepted	Passed
3	835,184.OC	Beijing Century GrandTech Corp., Ltd.	2015/12/24	Information transmission, Software and IT services	Guoyuan securities	Application accepted	Passed
4	830,815.OC	Beijing Blue Mountains Technology Co., Ltd.	2014/6/20	Manufacturing	Hualong securities	Application accepted	–
5	834,021.OC	Golden Times Culture Communications Corp., Ltd.	2015/10/30	Information transmission, Software and IT services	Tianfeng securities	Application accepted	Passed
6	836,263.OC	Beijing ZHTD Environmental Protection Technology Co., Ltd.	2016/3/22	Management of Water Conservancy, Environment and Public Facilities	China securities	Application accepted	Passed
7	430,090.OC	Tonghuijiashi (Beijing) Information Technology Co., Ltd.	2011/6/17	Information transmission, Software and IT services	Shengang securities	Application accepted	–
8	430,046.OC	Beijing SBR Information Technology Co., Ltd.	2009/2/18	Information transmission, Software and IT services	Western securities	Application accepted	Suspend
9	835,508.OC	Beijing In-To Digital Technology Stock Co., Ltd.	2016/1/15	Information transmission, Software and IT services	Wanlian securities	Application accepted	Passed
10	837,747.OC	Beijing Changjiang Culture Co., Ltd.	2016/6/22	Culture, Sports and Entertainment	Jinyuan securities	The record accepted	–

(continued)

Table 7.1 (continued)

Serial number	Stock code	Company name	Listing date	Category	Host broker	Preparation progress	Listing committee audit results
11	834,857.OC	AEPA&TSC architects engineers Inc.	2015/12/15	Scientific research and technology service	Guotai Junan	Application accepted	–
12	832,646.OC	Beijing Xunzhong Communication Technology Co., Ltd.	2015/6/16	Information transmission, Software and IT services	Nanjing securities	Filed	–
13	430,047.OC	Beijing Northland Biotech.Co., Ltd.	2009/2/18	Manufacturing	Zhongtai securities	Declared	–
14	837,592.OC	Hi-Think Yonder vision (Beijing)Tech.Co., Ltd.	2016/6/1	Information transmission, Software and IT services	Shen Wan Hongyuan	Filed	–
15	831,142.OC	Beijing Easted Information Technology Co., Ltd.	2014/9/11	Information transmission, Software and IT services	Changjiang securities	Filed	–
16	430,208.OC	Beijing Uxsino Software Co., Ltd.	2013/1/29	Information transmission, Software and IT services	Essence securities	Filed	–
17	836,346.OC	Beijing Emar Online Technology Co., Ltd.	2016/3/23	Information transmission, Software and IT services	Guosen securities	Filed	–
18	871,553.OC	BEIJING KAITENG JINGGONG GRAVORE CYLNDER Co., Ltd.	2017/5/17	Manufacturing	Caida securities	Filed	–
19	834,082.OC	China National Building Materials Technology Co., Ltd.	2015/11/5	Information transmission, Software and IT services	Cinda securities	Filed	–
20	430,014.OC	Beijing Hengye Century Technology Company Limited Co., Ltd.	2007/6/15	Manufacturing	Zheshang securities	Filed	–
21	837,344.OC	Beijing Tri-Prime Gene Pharmaceutical Co., Ltd.	2016/5/18	Manufacturing	Shen Wan Hongyuan	Filed	–
22	835,961.OC	Noble Family Wine and Liquor Co., Ltd.	2016/3/2	Wholesale and Retail businesses	CITIC securities	Filed	–

(continued)

Table 7.1 (continued)

Serial number	Stock code	Company name	Listing date	Category	Host broker	Preparation progress	Listing committee audit results
23	832,924.OC	Bright Stone Innovation Technology Group Co., Ltd.	2015/7/22	Manufacturing	Galaxy securities	Filed	–
24	836,208.OC	Greetec Co., Ltd.	2016/3/11	Scientific research and technology service	Minsheng securities	Filed	–
25	833,429.OC	Beijing Competitor Sports Science Technology Joint Stock Co., Ltd	2015/8/28	Manufacturing	Pacific securities	Filed	–

PS: "–" refers to the companies are not at the stage of list auditing by 1st July 2020, there is no result of auditing
Source Wind, Zhongguancun Listed Company Association

7.1.3 Zhongguancun NEEQ Listed Companies 'Select' Layer Matching Criterions Analysis

The 'Select' layer has set up four different criterions, including the profitability, growth ability, sales with research and market recognition with innovations. According to the *Measures for the hierarchical management of the national share transfer system for small and medium sized enterprises*, the listed companies has to meet the one of the following criterions when they choose to apply for public-offering and enter into 'Select' layer. 1. The total market value should not be below 200 million yuan, the net profit in each of the recent two years should not be below 15 million yuan and the weighted average return on net asset should not be below 8%. Otherwise, the net profit in the recent year should not be below 25 million and the weighted average return on net asset should not be below 8%. 2. The total market value should not be below 400 million, the average revenue in each of the recent two years should not be below 100 million and the growth on revenue in recent one year should not be below 30% with the positive net cash flow from operating activities. 3. The market value should not be below 800 million, the revenue in the recent year should not be below 200 million and the sum of R&D expenses divided by the sum of the revenue in the recent two years should not be below 8%. 4. The market value should not be below 1.5 billion and the sum of R&D expenses in recent two years should not be below 50 million. The market value addressed above is calculated by the directional offering price to the unspecified qualified investors. In addition, the *Layered management method* also requires the year-ended net asset value of listed companies should not be below 50 million when the public-offering and the entry of 'Select' layer are completed (Fig. 7.1).

Until 30th June 2020, there are 164 Zhongguancun companies listed in 'Innovation' according to the name list periodically sorted out by the NEEQ Co., Ltd in 2020.

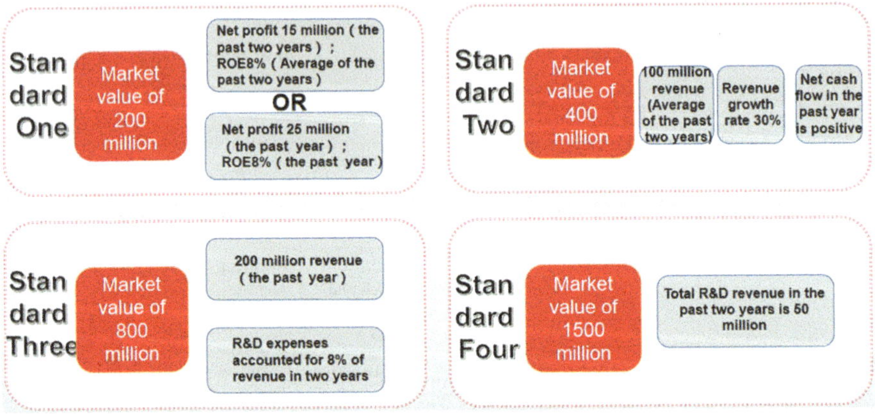

Fig. 7.1 The financial index of the NEEQ Select layer companies. *Source Wind*, compiled by ZLCA

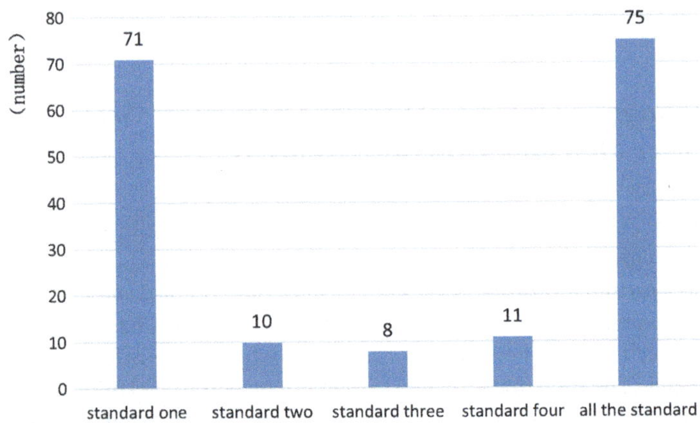

Fig. 7.2 The analysis of the matching standard of Zhongguancun listed companies for 'Select'. *Source Wind*, compiled by ZLCA

According to the 4 criterions of 'Select' layer mentioned above, there are 76 out of 164 Zhongguancun 'Innovation' layer companies have met the one of the 4 criterions to enter the 'Select' layer (Ding Ling technology is not qualified for 'Select' with respect to the requirement of year-ended net asset value). To sum up, there are 75 Zhongguancun NEEQ listed companies have met one of the 4 criterions to enter the 'Select' layer, accounts for 45.73%, which ranks 3rd in China (See Fig. 7.2) after Guang Dong province (77) and Jiang Su province (76). However, the R&D level of Beijing NEEQ companies takes the leading position in China. In the perspective of 3rd and 4th criteria which focus on the R&D level of companies, the number of qualified Beijing companies are far more than the other provinces (See Figs. 7.3 and 7.4).

Specifically, there are the number of 71 Zhongguancun NEEQ 'Innovation' listed companies satisfy the 1st criteria for the combined requirements of market value, net

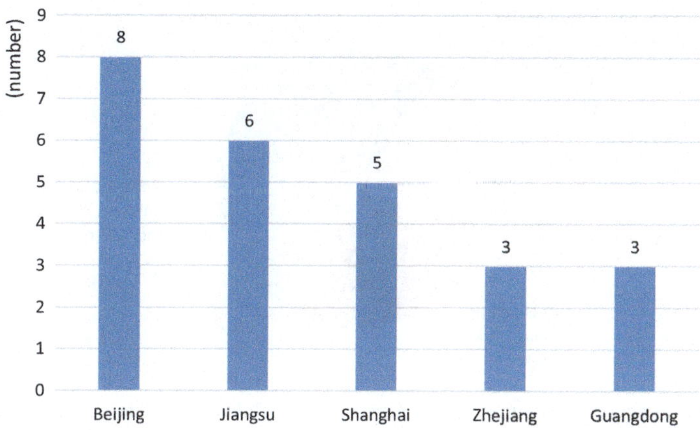

Fig. 7.3 The top 5 provinces which owns the largest number of companies satisfying the third criteria for 'Select' layer. *Source Wind*, compiled by ZLCA

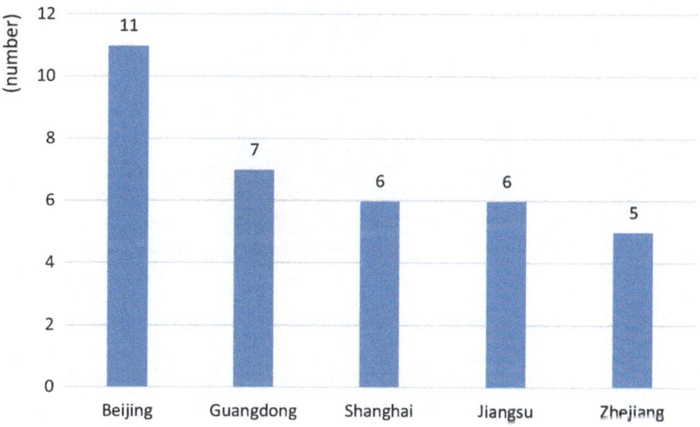

Fig. 7.4 The top 5 provinces which owns the largest number of companies satisfying the fourth criteria for 'Select' layer. *Source Wind*, compiled by ZLCA

profit and ROE. Additionally, there are 10 companies satisfy the 2nd criteria for the requirements of market value, revenue, growth of revenue and cash-flow. Furthermore, there are 8 companies satisfy the 3rd criteria for the requirement of market value, revenue and R&D intensity. Eventually, there are 11 companies satisfy the 4th criteria for the requirements of market value and R&D expenses. Overall, the number of Zhongguancun NEEQ listed companies which satisfy the 1st criteria is more than the others as the 13 companies both satisfy the two of market value and financial standards. There are 6 companies satisfy the three of the market value and financial standards.

From the perspective of the industry, many companies have met the 'Select' layer criterions, including 32 Information transmission, Software and IT services companies; 21 Manufacturing companies, 6 Leasing and Business services companies, 4

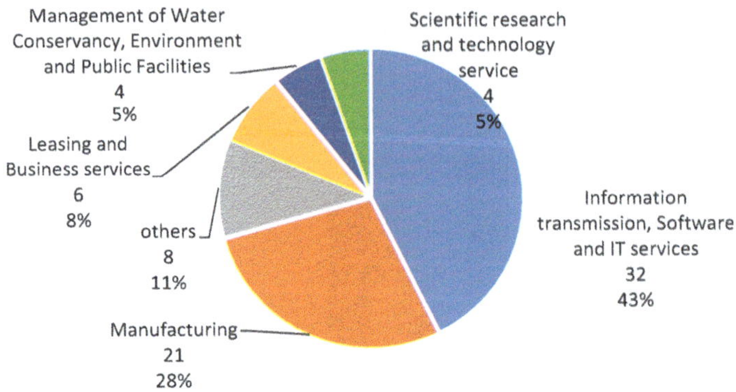

Fig. 7.5 The industry distribution of Zhongguancun candidates to 'Select' layer. *Source Wind*, compiled by ZLCA

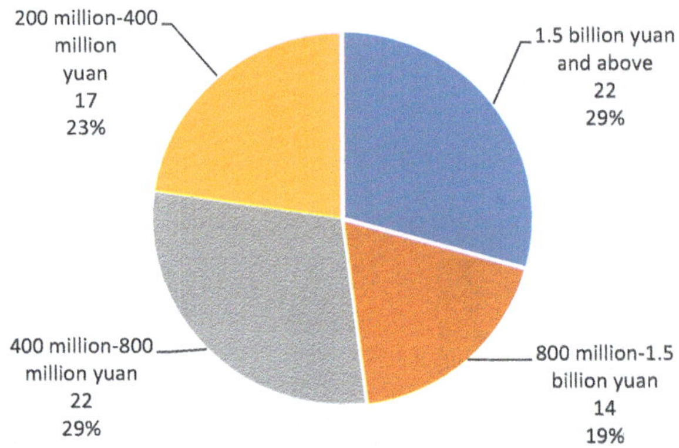

Fig. 7.6 The market value distribution of Zhongguancun candidates to 'Select' layer. *Source Wind*, compiled by ZLCA

Management of Water Conservancy, Environment and Public Facilities companies, 4 Scientific research and technology service companies and 8 companies in other industries. The majority of them are concentrated in the information and technology innovative industries, the most of the candidates of 'Select' layer companies are the leaders of different sub-industries (see Fig. 7.5).

From the perspective of market value, the average market value of the 75 Zhong Guan Cun candidates of the 'Select' layer is 1.455 billion yuan. There are 22 companies valued more than 1.5 billion, accounting for 29%; 14 companies valued between 800 million to 1.5 billion, accounting for 19%; 22 companies valued between 400 to 800 million, accounting for 29%; 17 companies valued between 200 to 400 million, accounting for 23% (see Fig. 7.6).

From the perspective of revenue in 2019, the average revenue of 75 Zhong Guan Cun candidates of 'Select' layer is 1.042 billion yuan. There are 3 companies receive the revenue more than 5 billion yuan in 2019, accounting for 4%; 7 companies receive the revenue between 1 to 5 billion yuan in 2019, accounting for 9%; 15 companies

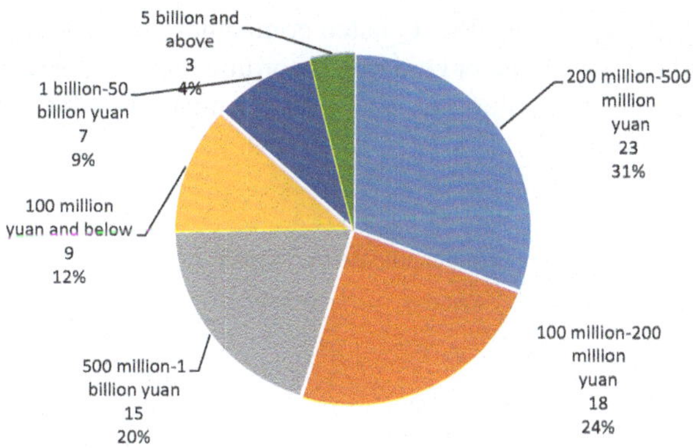

Fig. 7.7 The revenue distribution of Zhongguancun candidates to 'Select' layer in 2019. *Source Wind*, compiled by ZLCA

receive the revenue between 500 million to 1 billion in 2019, accounting for 20%; 23 companies receive the revenue between 200 to 500 million in 2019, accounting for 31%; 18 companies receive the revenue between 100 to 2 million in 2019, accounting for 24%; 9 companies receive the revenue less than 100 million in 2019, accounting for 12% (see Fig. 7.7).

From the perspective of net profit in 2019, the average net profit of 75 Zhong Guan Cun candidates of 'Select' layer is 60.3805 million yuan. There are 9 companies receive the net profit more than 100 million yuan in 2019, accounting for 12%; 16 companies receive the net profit between 50 to 100 million in 2019, accounting for 21%; 13 companies receive the net profit between 35 to 50 million in 2019, accounting for 17%; 21 companies receive the net profit between 25 to 35 million in 2019, accounting for 28%; 14 companies receive the net profit between 0 to 25 million in 2019, accounting for 19%; 2 companies have negative net profit in 2019, accounting for 3% (see Fig. 7.8).

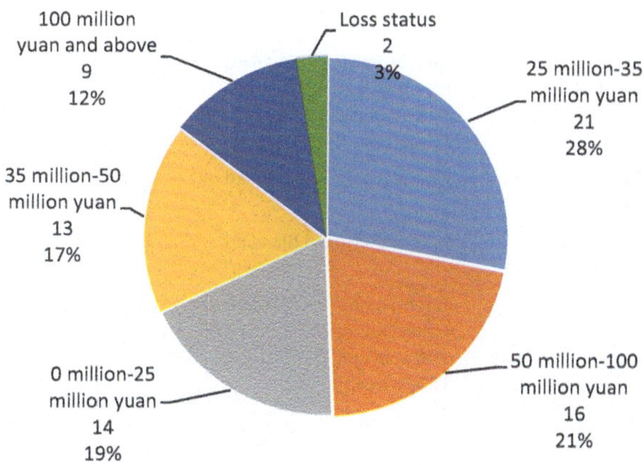

Fig. 7.8 The net profit distribution of Zhongguancun candidates to 'Select' layer in 2019. *Source Wind*, compiled by ZLCA

From the 75 Zhongguancun NEEQ listed companies above, there are 25 companies have applied for 'Select' layer guidance recording, the other 50 companies which satisfy at least one of the 4 criterions. The specific situation of the 50 companies are shown following (Table 7.2):

Table 7.2 The status of Zhongguancun candidate for 'Select' layer

	Stock code	Company Name	Category	Standard	Market value as of June 30, 2020 (100 million yuan)	Net profit by 2019 (ten thousand yuan)	revenue to 2019 (ten thousand yuan)
1	430,005.OC	HTA Co., Ltd.	Manufacturing	Standard one/four	83.87	27,925.70	113,618.77
2	430,037.OC	Beijing Unifly Scientific and Technology Company Limited	Manufacturing	Standard one	3.13	3724.12	43,247.66
3	430,074.OC	Golden Spring Internet of Things Inc.	Manufacturing	Standard one	13.95	7369.14	55,834.64
4	430,109.OC	Beijing CIC Technology Co., Ltd.	Manufacturing	Standard one	2.24	2150.20	18,109.45
5	430,193.OC	Win Media Network Technology Co., Ltd.	Information transmission, Software and IT services	Standard one	10.31	9609.24	46,543.57
6	430,277.OC	Beijing Shengshang Education & Tech Co., Ltd.	Leasing and Business services	Standard one/two	59.70	17,339.49	75,283.27
7	430,287.OC	Beijing Kingpeng Global Husbandry Technology Co., Ltd.	Manufacturing	Standard one	3.86	2535.07	70,257.02
8	430,290.OC	Beijing HeroopSys Co., Ltd.	Information transmission, Software and IT services	Standard one	5.28	1741.98	6228.41
9	831,222.OC	Beijing Intolo Decoration Co., Ltd.	Construction	Standard one	2.29	4286.81	138,245.43
10	831,344.OC	Ficont Industry (Beijing) Co., Ltd.	Manufacturing	Standard one/two	12.44	14,177.92	53,907.70
11	831,727.OC	Beijing Steel Information Network Co., Ltd.	Information transmission, Software and IT services	Standard one	3.32	3503.10	495,213.00

(continued)

Table 7.2 (continued)

	Stock code	Company Name	Category	Standard	Market value as of June 30, 2020 (100 million yuan)	Net profit by 2019 (ten thousand yuan)	revenue to 2019 (ten thousand yuan)
12	832,145.OC	Beijing Henghe Information & Technology Co., Ltd.	Management of Water Conservancy, Environment and Public Facilities	Standard one	7.19	3203.67	12,571.78
13	832,694.OC	Beijing Wingain Machinery and Electronics Co., Ltd.	Manufacturing	Standard one	3.26	2538.71	50,778.89
14	833,024.OC	Beijing XinZhiHeng Techology Co., Ltd.	Information transmission, Software and IT services	Standard one/two	16.37	5666.90	51,992.42
15	833,629.OC	HOLLYCRM Co., Ltd.	Information transmission, Software and IT services	Standard one/two	5.37	3257.66	24,110.83
16	833,741.OC	Country Holidays (Beijing) International Travel Service Incorporated Co., Ltd.	Leasing and Business services	Standard one	7.96	3670.62	200,216.50
17	833,966.OC	Guodiankangneng Technology Stock Co., Ltd.	Scientific research and technology service	Standard one	12.25	10,565.81	33,441.42
18	834,037.OC	Beijing Dragon Century Technology Inc.	Information transmission, Software and IT services	Standard one	9.59	3726.01	90,733.83
19	834,063.OC	Beijing Truck Information Technology Co., Ltd.	Information transmission, Software and IT services	Standard one	6.27	2988.68	15,710.05
20	834,218.OC	Hecom (Beijing) Technology Co., Ltd.	Information transmission, Software and IT services	Standard one	18.00	-3143.91	12,718.30
21	834,327.OC	Beijing Chexun Internet Co., Ltd.	Information transmission, Software and IT services	Standard one	3.14	2617.56	23,335.61
22	834,476.OC	Wishart Media Co., Ltd.	Leasing and Business services	Standard one	5.66	1778.54	10,832.94

(continued)

Table 7.2 (continued)

	Stock code	Company Name	Category	Standard	Market value as of June 30, 2020 (100 million yuan)	Net profit by 2019 (ten thousand yuan)	revenue to 2019 (ten thousand yuan)
23	834,785.OC	Beijing Yunchang Game Technology Co., Ltd.	Information transmission, Software and IT services	Standard one	6.35	5566.80	24,407.08
24	834,802.OC	B&G Holding Company	Information transmission, Software and IT services	Standard one	33.44	2288.04	45,082.71
25	835,281.OC	Highly Information Industry Co., Ltd.	Wholesale and Retail businesses	Standard one	25.21	21,560.43	2,083,561.75
26	835,291.OC	Beijing Holyzone Technology Corporation	Information transmission, Software and IT services	Standard one/two	4.41	2682.53	49,663.17
27	835,363.OC	Beijing Suntray Technology Corporation	Information transmission, Software and IT services	Standard one	3.61	2244.55	15,769.05
28	835,911.OC	Beijing Agrichina Pharmaceutical Co., Ltd.	Manufacturing	Standard one	4.96	2099.56	21,508.95
29	836,348.OC	Beijing Huiheng Environmental Protection Engineering Co., Ltd.	Management of Water Conservancy, Environment and Public Facilities	Standard one	2.31	1572.51	9255.23
30	836,464.OC	Beijing Huachengzhiyun Soft Co., Ltd.	Information transmission, Software and IT services	Standard one	2.71	2507.81	11,764.64
31	836,825.OC	China Bidding Co., Ltd.	Information transmission, Software and IT services	Standard one	4.15	1943.29	14,804.87
32	836,870.OC	Beijing Sunway Technology Co., Ltd.	Information transmission, Software and IT services	Standard one	3.73	2609.15	9397.19
33	836,885.OC	Beijing Headspring Technology Co., Ltd	Information transmission, Software and IT services	Standard one	3.77	2439.43	14,212.25

(continued)

Table 7.2 (continued)

	Stock code	Company Name	Category	Standard	Market value as of June 30, 2020 (100 million yuan)	Net profit by 2019 (ten thousand yuan)	revenue to 2019 (ten thousand yuan)
34	837,840.OC	China Electronic Cowan Science & Technology Co., Ltd.	Information transmission, Software and IT services	Standard one	9.72	1975.44	7938.03
35	837,899.OC	Beijing Orient Tonghua Technology Co., Ltd.	Electricity, Heat, Gas and Water production and supply industry	Standard one	8.82	5973.19	19,252.87
36	837,950.OC	Beijing Action Infornation Technology Co., Ltd.	Information transmission, Software and IT services	Standard one	4.77	2191.96	34,431.96
37	838,168.OC	Beijing Kuaiyu Electronics Co., Ltd.	Manufacturing	standard one	4.54	3039.40	15,490.37
38	838,334.OC	Beijing JZHT Finance Securities Service Co., Ltd.	Leasing and Business services	Standard one	6.62	3518.46	7585.22
39	838,740.OC	Beijing BangYuan environment science and technolgy limited Co., Ltd.	Management of Water Conservancy, Environment and Public Facilities	Standard one	3.85	2762.59	9053.20
40	839,463.OC	Beijing SG-Culture Media Corp., Ltd.	Culture, Sports and Entertainment	Standard one	4.69	3079.99	16,770.98
41	839,483.OC	Yonyou Fintech Technology Co., Ltd.	Information transmission, Software and IT services	Standard one/three/four	18.77	6137.71	38,069.26
42	839,622.OC	Trustwin Communication Brand Consulting(Beijing) Co., Ltd.	Leasing and Business services	Standard one	3.71	2750.62	33,403.81
43	839,994.OC	Know-how Education Ltd.	Education	Standard one	4.96	1760.29	9079.49
44	870,381.OC	Beijing 797 Audio Co., Ltd.	Manufacturing	Standard one	7.15	6275.05	30,622.26
45	870,992.OC	Beijing ZBX Information Technology Co.,Ltd.	Information transmission, Software and IT services	Standard one	4.21	2274.90	13,135.18
46	872,808.OC	Sugon Data Energy (Beijing) Co., Ltd.	Manufacturing	Standard one/two/three	12.55	2819.81	29,170.87

(continued)

Table 7.2 (continued)

	Stock code	Company Name	Category	Standard	Market value as of June 30, 2020 (100 million yuan)	Net profit by 2019 (ten thousand yuan)	revenue to 2019 (ten thousand yuan)
47	430,034.OC	Beijing Jiuzhoudadi Bio-Tech Group Co., Ltd.	Manufacturing	Standard one	16.04	8088.26	165,309.93
48	831,527.OC	MetaSpace (Beijing) Air Dome Corp.	Construction	Standard one	2.76	2573.93	18,789.20
49	835,179.OC	Beijing KaiDe Quartz Co., Ltd.	Manufacturing	Standard one/two	15.93	3246.04	15,450.48
50	837,940.OC	Brand Union (Beijing) Consulting Co., Ltd.	Leasing and Business services	Standard one	2.43	2512.85	16,401.96

Source Wind, compiled by ZLCA

7.2 The Introduction of the Key Companies to 'Select' Layer

7.2.1 HTA Co., Ltd.

1. Basic Information of the Company

HTA Co., Ltd. (hereinafter referred to as "HTA") was established in 2001, publicly listed in NEEQ (National Equities Exchange and Quotations) in 2006 (stock code: 430,005.OC). The Market value of the HTA is 5.037 billion *yuan* on 31st of December 2019.

HTA is the largest Radioactive Isotope Research & Development center, manufacturer and supplier in China. The main source of revenue was composed of the medicine for In-vivo diagnosis, Therapeutic radiopharmaceuticals, In-vitro immunoassay kit, Radioactive source, Radioactive medical devices, Radiolabeled compound, Tracer, radioactive reference materials and more than 70 different types of isotopes and 300 types of products.

The development of Nuclear technology application industry highly correlates with the rise of macro-economy of the state and living standard. The government has introduced a series of related policies and regulations to promote the development of healthcare industry, the Nuclear technology application industry opportunities therefore can be expanded toward healthcare industry. Currently, the volume of Nuclear technology application industry is fairly low, there is much space for it to develop with respect to the improvement in the range of application, technology level, application environment and market maturity.

The customer concentration is significantly low, the ratios as the each of top 5 customers sales by the total revenue are 3.55, 1.48, 1.16, 1.04, 0.98%, totaling 8.21%. The company doesn't reveal the specific names of the top 5 customers in 2019.

2. Financial Analysis

(1) Basic Financial Data (Table 7.3)

Table 7.3 Basic financial data for HTA from 2017 to 2019 (unit (*yuan* in 10 thousand))

	2019	2018	2017
Total assets	217,178.15	166,155.16	135,779.59
Net income attributed to shareholders	110,107.17	90,860.94	73,324.22
Revenue	113,618.77	99,620.10	85,725.99
Net profit	27,925.70	25,364.93	21,618.38
Cash flow from operating activities	19,878.05	15,555.69	18,381.53

Source Wind, compiled by ZLCA

(2) Primary Financial Data (Table 7.4)

Table 7.4 Primary financial data for HTA from 2017 to 2019

	2019	2018	2017
Gross profit margin (%)	62.38	64.55	64.06
Net profit margin (%)	24.58	25.46	25.22
Debt ratio (%)	39.17	33.13	33.94
Current ratio	1.67	2.08	2.24
Quick ratio	1.44	1.79	1.98
Accounts receivable turnover ratio	2.37	2.48	2.49

Source Wind, compiled by ZLCA

(3) Income Structure (Table 7.5)

Table 7.5 The income structure for HTA from 2017 to 2019 (unit *yuan* in ten thousands, %)

Category	2019		2018		2017	
	Amount	Ratio	Amount	Ratio	Amount	Ratio
Radiopharmaceuticals and services	98,166.29	86.40	84,507.00	84.83	74,498.11	86.90
Radioactive source income	14,304.45	12.59	14,366.41	14.42	9449.72	11.02
Radiation income	1148.03	1.01	746.68	0.75	1778.16	2.07
Total	113,618.77	100.00	99,620.10	100.00	85,725.99	100.00

Source Wind, compiled by ZLCA

7.2.2 *Beijing Unifly Scientific and Technology Company Limited*

1. Basic Information of the Company

Beijing Unifly Scientific and Technology Company Limited (hereinafter referred to as 'Unifly') was established in 1995, publicly listed in NEEQ (National Equities Exchange and Quotations) in 2008 (stock code: 430,037.OC). The Market value of the Unifly is 255 million *yuan* on 31st of December 2019.

The company is a high-tech enterprise that develops technology in high-tech fields to produce and sell high-tech products. The main business is to apply new materials technology to the research, development, production and sales of energy-saving and environmental-friendly fast-moving consumer goods for vehicles. The company owns the independently researched and developed patented technology of inorganic nonmetallic ceramic functional materials, and combines it with traditional fast-moving consumer goods such as motor vehicle filters and lubricants. The company successfully develops and transforms a variety of series products such as environmental protection and energy-saving filters, long-acting low-carbon lubricants, low-tension environmental protection glass cleaning agents, antifreeze coolant, etc.

The market demand of China's auto supplies industry is continuously increasing. Although automobile production and sales continued to grow negatively, the demand for environmental protection products surged, and the demand for energy-saving and environmental protection products under the guidance of policies was obvious. The production and sales volume of the front car market decreased, while the quantity of the rear car market continued to grow. The domestic auto insurance industry is developing rapidly, and the volume of auto insurance business is massive.

The customer concentration is fairly high, the ratios as the each of top 5 customers sales by the total revenue are 10.66, 6.03, 5.68, 5.42, 3.74%, totaling 31.53%.

2. Financial Analysis

   (1) Basic Financial Data (Table 7.6)

Table 7.6 Basic financial data for HTA from 2017 to 2019 (unit (*yuan* in 10 thousand))

	2019	2018	2017
Total assets	54,531.98	47,786.15	49,919.30
Net income attributed to shareholders	39,436.61	35,158.23	37,005.04
Revenue	43,247.66	30,775.78	28,533.16
Net profit	3724.12	2272.47	2593.24
Cash flow from operating activities	763.16	1355.39	-2963.98

Source Wind, compiled by ZLCA

(2) The Primary Financial Data (Table 7.7)

Table 7.7 Primary financial data for Unifly from 2017 to 2019

	2019	2018	2017
Gross profit margin (%)	25.92	25.83	22.50
Net profit margin (%)	8.61	7.38	9.09
Debt ratio (%)	25.99	24.68	24.26
Current ratio	3.26	3.16	3.02
Quick ratio	1.77	1.79	2.16
Accounts receivable turnover ratio	2.28	1.62	1.66

Source Wind, compiled by ZLCA

(3) Income Structure (Table 7.8)

Table 7.8 The income structure for Unifly from 2017 to 2019 (unit *yuan* in ten thousands, %)

Category	2019		2018		2017	
	Amount	Ratio	Amount	Ratio	Amount	Ratio
Long-acting low-carbon lubricating oil	23,828.27	55.35	22,611.62	74.53	22,710.46	80.35
New functional material products	7649.76	17.77	622.88	2.05	–	
Low tension curing liquid products	4724.09	10.97	443.34	1.46	478.49	1.69
Commission service income	3154.32	7.33	3104.15	10.23	493.48	1.75
Post-market service income	2539.64	5.90	1097.11	3.62	–	–
Energy-saving and environment-friendly filters and supporting components	1157.58	2.68	2457.86	8.11	4484.14	15.87
Other classes	–	–	–	–	92.25	0.34
Total	43,053.66	100.00	30,334.96	100.00	28,262.82	100.00

Source Wind, compiled by ZLCA

7.2.3 *Golden Spring Internet of Things Inc*

1. Basic Information of the Company

Golden Spring Internet of Things Inc (hereinafter referred to as 'Golden Spring Internet of Things') was established in 2004, publicly listed in NEEQ (National Equities Exchange and Quotations) in 2010 (stock code: 430,074.OC). The Market value of the Golden Spring Internet of Things is 1.534 billion *yuan* on 31st of December 2019.

The company is a comprehensive solution-provider for RFID production and application in the Internet of Things, and a high-tech enterprise specializing in RFID (including smart tags, contactless smart cards and composite cards) production and personalized comprehensive solutions. The main products are non-contact smart cards, dual-interface smart cards, production equipment packaged with smart labels and comprehensive solutions for production services; Radio frequency identification reading and writing equipment and its application comprehensive solution.

The internet of things industry in China is developing at a high speed driven by the policy and the demand of downstream industries. The internet of things is a national strategic emerging industry with great market development potential and increasingly tremendous market competition. The both domestic and foreign manufacturers, which are mainly engaged in low-end RFID packaging equipment in the domestic packaging field gradually grow and develop through fierce competition, strengthen R&D and innovation, and launch high-performance products with independent intellectual property rights, which will inevitably intensify the market competition in the field of RFID high-end packaging equipment.

The customer concentration is fairly high, the ratios as the each of top 5 customers sales by the total revenue are 14.66, 12.67, 11.49, 7.47, 5.04%, totaling 51.33%.

2. Financial Analysis

(1) Basic Financial Data (Table 7.9)

Table 7.9 Basic financial data for golden spring internet of things from 2017 to 2019 (unit (*yuan* in 10 thousand))

	2019	2018	2017
Total assets	121,887.38	89,471.38	72,625.19
Net income attributed to shareholders	75,760.82	50,560.83	44,432.34
Revenue	55,834.64	50,812.44	41,366.92
Net profit	7369.14	6035.29	5073.63
Cash flow from operating activities	−1636.39	1082.88	−2264.22

Source Wind, compiled by ZLCA

(2) Primary Financial Data (Table 7.10)

Table 7.10 Primary financial data for golden spring internet of things from 2017 to 2019

	2019	2018	2017
Gross profit margin (%)	28.39	23.34	27.77
Net profit margin (%)	13.20	11.88	12.27
Debt ratio (%)	37.47	43.13	38.25
Current ratio	2.48	2.13	2.12
Quick ratio	1.80	1.21	1.20
Accounts receivable turnover ratio	8.24	6.60	5.23

Source Wind, compiled by ZLCA

(3) Income Structure (Table 7.11)

Table 7.11 The income structure for golden spring internet of things from 2017 to 2019 (unit *yuan* in ten thousands, %)

Category	2019		2018		2017	
	Amount	Ratio	Amount	Ratio	Amount	Ratio
RFID application system	34,197.12	61.25	6087.77	11.98	18,029.00	43.58
RFID reader	10,143.89	18.17	1590.61	3.13	7,784.95	18.82
RFID intelligent production equipment	8482.91	15.19	819.24	1.61	8,714.48	21.07
Raw materials and spare parts	2459.13	4.40	15,858.67	31.21	3,607.85	8.72
Production technology service	551.58	0.99	26,456.14	52.07	3,230.64	7.81
Total	55,834.63	100.00	50,812.43	100.00	41,366.92	100.00

Source Wind, compiled by ZLCA

7.2.4 Beijing CIC Technology CO., Ltd.

1. Basic Information of the Company

Beijing CIC Technology CO., Ltd. (hereinafter referred to as 'CIC') was established in 2007, publicly listed in NEEQ (National Equities Exchange and Quotations) in 2012 (stock code: 430,109.OC). The Market value of the CIC is 192 million *yuan* on 31st of December 2019.

The company's main business is to provide integrated solutions for mobile video information collection, transmission and processing for industry customers in urban operation and management, and to provide intelligent bus networking services. The main customers are local bus companies, which can obtain revenue and market share by providing bus companies with "software + hardware" products that meet the requirements of bus informationization and intelligence.

The company belongs to the intelligent public transport industry. The Software and hardware are combined in the intelligent public transport industry, with software as the mainstay and hardware as the supplement. The product modularization, customization and matching degrees are high. There is a high market volume for the product development. The problems of urban traffic congestion, lack of public transport capacity and surge in traffic volume are prominent with the continuous development and expansion of cities. The local government are expecting to build intelligent transportation through the internet to solve urban traffic problems. The urban public transport industry in China is in an accelerated development stage.

The customer concentration is fairly high, the ratios as the each of top 5 customers sales by the total revenue are 27.04, 11.80, 9.90, 5.60, 5.27%, totaling 59.61%.

2. Financial Analysis

(1) Basic Financial Data (Table 7.12)

Table 7.12 Basic financial data for CIC from 2017 to 2019 (unit (*yuan* in 10 thousand))

	2019	2018	2017
Total assets	23,503.23	18,039.74	19,873.31
Net income attributed to shareholders	16,261.40	14,744.23	13,205.37
Revenue	18,109.45	11,822.80	10,784.93
Net profit	2150.20	2077.49	1757.66
Cash flow from operating activities	−996.99	−1750.21	−983.09

Source Wind, compiled by ZLCA

(2) Primary Financial Data (Table 7.13)

Table 7.13 Primary financial data for CIC from 2017 to 2019

	2019	2018	2017
Gross profit margin (%)	41.41	50.08	50.74
Net profit margin (%)	11.87	17.57	16.30
Debt ratio (%)	29.57	18.08	33.61
Current ratio	3.11	5.11	2.76
Quick ratio	2.92	4.47	2.46
Accounts receivable turnover ratio	1.92	1.70	1.92

Source Wind, compiled by ZLCA

(3) Income Structure (Table 7.14)

Table 7.14 The income structure for CIC from 2017 to 2019 (unit *yuan* in ten thousands, %)

Category	2019		2018		2017	
	Amount	Ratio	Amount	Ratio	Amount	Ratio
On-board hardware products	15,635.91	86.34	9230.06	78.07	9965.69	92.40
On-board software products	1610.73	8.89	1734.05	14.67	556.59	5.16
Technology development	385.12	2.13	134.45	1.14	225.48	2.09
Technology services	340.87	1.88	435.14	3.68	21.20	0.20
Advertising service	136.81	0.76	36.83	0.31	15.97	0.15
Project income	–	–	252.27	2.13	–	–
Total	18,109.44	100.00	11,822.80	100.00	10,784.93	100.00

Source Wind, compiled by ZLCA

7.2.5 *Win Media Network Technology Co., Ltd.*

1. Basic Information of the Company

Win Media Network Technology Co., Ltd. (hereinafter referred to as 'Win Media') was established in 2003, publicly listed in NEEQ (National Equities Exchange and Quotations) in 2012 (stock code: 430,193.OC). The Market value of the Win Media is 270 million *yuan* on 31st of December 2019.

The company's main business is online self-media marketing and online advertising business. The company provides high-quality video content production, distribution and promotion services for customers such as game platform, e-commerce platform and network service platform on carriers such as short video self-media and social self-media by the self-developed platform, the years of data accumulation from self-media and short video promotion, as well as service advantages in multi-platform content production, dissemination and drainage, and combining the professional marketing strategy of MCN service team.

The company belongs to the internet and related service industries, and the market is generally optimistic about the internet industry. More and more enterprises are entering the internet field, and the competition is greater. In the field of self-media marketing, the user base of self-media platform continues to expand, and the industry develops and updates rapidly.

The customer concentration is fairly high, the ratios as the each of top 5 customers sales by the total revenue are 17.40, 10.57, 6.23, 4.27, 3.59%, totaling 42.06%. The company doesn't reveal the specific names of the top 5 customers in 2019.

2. Financial Analysis

(1) Basic Financial Data (Table 7.15)

Table 7.15 Basic financial data for win media from 2017 to 2019 (unit (*yuan* in 10 thousand))

	2019	2018	2017
Total assets	57,812.02	50,163.35	36,932.78
Net income attributed to shareholders	53,647.52	44,038.28	26,069.73
Revenue	46,543.57	41,031.54	34,782.90
Net profit	9609.24	10,761.98	8327.32
Cash flow from operating activities	1477.05	3424.97	6100.92

Source Wind, compiled by ZLCA

(2) Primary Financial Data (Table 7.16)

Table 7.16 Primary financial data for win media from 2017 to 2019

	2019	2018	2017
Gross profit margin (%)	25.41	31.60	31.69
Net profit margin (%)	20.65	26.23	23.94
Debt ratio (%)	7.20	12.21	29.41
Current ratio	8.24	6.15	1.80
Quick ratio	8.24	6.15	1.80
Accounts Receivable Turnover Ratio	3.39	3.66	6.56

Source Wind, compiled by ZLCA

(3) Income Structure (Table 7.17)

Table 7.17 The income structure for win media from 2017 to 2019 (unit *yuan* in ten thousands, %)

Category	2019		2018		2017	
	Amount	Ratio	Amount	Ratio	Amount	Ratio
Information marketing	46,543.57	100.00	–	–	–	–
Technical service	–	–	37,014.13	90.21	29,171.45	83.87
Network marketing	–	–	3293.82	8.03	5611.45	16.13
Other business	–	–	723.58	1.76	–	–
Total	46,543.57	100.00	41,031.53	100.00	34,782.90	100.00

Source Wind, compiled by ZLCA

7.2.6 *Beijing Shengshang Education and Tech Co., Ltd.*

1. Basic Information of the Company

Beijing Shengshang Education & Tech Co., Ltd. (hereinafter referred to as 'Shengshang Education') was established in 2005, publicly listed in NEEQ (National Equities Exchange and Quotations) in 2013 (stock code: 430,277.OC). The Market value of the Shengshang Education is 5.105 billion *yuan* on 31st of December 2019.

The company provides consulting services related to enterprise management and strategy, finance and capital to customers through internal or external hired experts and tutors, and obtains consulting service income. The company's customers are scattered and diversified, mainly concentrated in small, medium-sized and micro enterprises, including business owners and business teams with needs in internal management, strategy, capital, investment, IPO and listing of the NEEQ.

The consulting service industry where the company is involved is highly competitive, with numerous consulting service companies and different service quality. The competition of traditional consulting service products on the internet is fierce, and online education products are becoming intelligent. Traditional technology products will be replaced or eliminated in the future. The Company needs to build core products that integrate the online and offline competitive advantages in order to adapt to the severe market competition environment.

The customer concentration is significantly low, the ratios as the each of top 5 customers sales by the total revenue are 0.22, 0.18, 0.16, 0.15, 0.14%, totaling 0.84%.

2. Financial Analysis

(1) Basic Financial Data (Table 7.18)

Table 7.18 Basic financial data for Shengshang education from 2017 to 2019 (unit (*yuan* in 10 thousand))

	2019	2018	2017
Total assets	34,198.39	34,200.63	17,397.02
Net income attributed to shareholders	18,199.97	12,884.03	3174.71
Revenue	75,283.27	52,021.43	8738.90
Net profit	17,339.49	9725.14	477.16
Cash flow from operating activities	7829.84	11,928.68	12,130.65

Source Wind, compiled by ZLCA

(2) Primary Financial Data (Table 7.19)

Table 7.19 Primary financial data for Shengshang education from 2017 to 2019

	2019	2018	2017
Gross profit margin (%)	46.08	41.60	45.56
Net profit margin (%)	23.03	18.69	5.46
Debt ratio (%)	46.78	62.28	81.75
Current ratio	0.87	0.65	1.21
Quick ratio	0.87	0.65	1.18
Accounts receivable turnover ratio	0.00	0.00	12.29

Source Wind, compiled by ZLCA

(3) Income Structure (Table 7.20)

Table 7.20 The income structure for Shengshang education from 2017 to 2019 (unit *yuan* in ten thousands, %)

Category	2019		2018		2017	
	Amount	Ratio	Amount	Ratio	Amount	Ratio
Income from consulting services	75,283.27	100.00	47,621.39	91.54	3556.39	40.70
Vitamin premix feed	–	–	4143.66	7.97	4428.55	50.68
Animal health products	–	–	241.20	0.46	753.96	8.62
Other business	–	–	15.18	0.03	–	–
Total	75,283.27	100.00	52,021.43	100.00	8738.90	100.00

Source Wind, compiled by ZLCA

7.2.7 Beijing Kingpeng Global Husbandry Technology Co., Ltd.

1. Basic Information of the Company

Beijing Kingpeng Global Husbandry Technology Co., Ltd. (hereinafter referred to as 'Kingpeng Global Husbandry') was established in 2006, publicly listed in NEEQ (National Equities Exchange and Quotations) in 2013 (stock code: 430,287.OC). The Market value of the Kingpeng Global Husbandry is 63 million *yuan* on 31st of December 2019.

The company's main business is to provide customers in the livestock industry with overall solutions for large-scale livestock farm farming, as well as R&D, production, sales, installation and related technical training services for supporting farming equipment. The main products are cattle farm business: milking equipment, manure treatment system, environmental control equipment and pasture management software. Pig farm business: other farm equipment, automatic feeding equipment, intelligent sow feeding station (ESF), environmental control equipment, manure treatment system and other farm equipment.

Jingpeng Huanyu animal husbandry turnkey engineering technology has strong advantages in the industry and has been widely recognized by the market. To form the EPC engineering technology mode of combining animal husbandry and breeding around the world with manure energy utilization, energy saving and emission reduction, breeding with species, and planting as the "consumption pool" of livestock manure, so as to realize the "animal husbandry" function of turnkey engineering technology and leap to the "breeding cycle ecology" function. Moreover, it has a complete turnkey engineering qualification. The main business of competitors is selling equipment, while some competitors are left behind in turnkey engineering technology and lack of turnkey engineering qualification.

The customer concentration is fairly high, the ratios as the each of top 5 customers sales by the total revenue are 22.15, 11.30, 5.63, 4.30, 3.97%, totaling 47.35%.

2. Financial Analysis

(1) Basic Financial Data (Table 7.21)

Table 7.21 Basic financial data for Kingpeng global husbandry from 2017 to 2019 (unit (*yuan* in 10 thousand))

	2019	2018	2017
Total assets	52,931.81	42,675.94	33,053.02
Net income attributed to shareholders	18,162.00	15,626.93	13,692.74
Revenue	70,257.02	54,467.93	36,659.93
Net profit	2535.07	2249.19	2455.36
Cash flow from operating activities	−3933.09	2505.37	−1203.69

Source Wind, compiled by ZLCA

(2) Primary Financial Data (Table 7.22)

Table 7.22 Primary financial data for Kingpeng global husbandry from 2017 to 2019

	2019	2018	2017
Gross profit margin (%)	13.88	15.53	20.29
Net profit margin (%)	3.61	4.13	6.70
Debt ratio (%)	65.69	63.38	58.57
Current ratio	1.48	1.53	1.65
Quick ratio	0.95	0.99	1.25
Accounts receivable turnover ratio	5.17	3.19	2.14

Source Wind, compiled by ZLCA

(3) Income Structure (Table 7.23)

Table 7.23 The income structure for Kingpeng global husbandry from 2017 to 2019 (unit *yuan* in ten thousands, %)

Category	2019		2018		2017	
	Amount	Ratio	Amount	Ratio	Amount	Ratio
Pig industry	44,392.10	63.19	31,364.44	57.58	22,345.41	60.95
Cattle rearing	25,832.56	36.77	23,103.49	42.42	14,314.51	39.05
Poultry industry	32.36	0.04	0.00	0.00	–	–
Total	70,257.02	100.00	54,467.93	100.00	36,659.92	100.00

Source Wind, compiled by ZLCA

7.2.8 Beijing HeroopSys Co., Ltd.

1. Basic Information of the Company

Beijing HeroopSys Co., Ltd. (hereinafter referred to as 'HeroopSys') was established in 2004, publicly listed in NEEQ (National Equities Exchange and Quotations) in 2013 (stock code: 430,290.OC). The Market value of the HeroopSys is 710 million *yuan* on 31st of December 2019.

The company is a service provider of intelligent industrial manufacturing and intelligent optimization solutions. The company's main business is the development and service of soft technologies such as process optimization technology, artificial intelligence technology, multi-device and large-scale system coordination optimization technology, enterprise-level big data mining technology, industrial internet service platform and intelligent optimization process simulation software.

"Intelligence plus", "internet plus", energy saving and consumption reduction will be the main themes of various industries in China. The new generation information technology industry with core technologies such as artificial intelligence, big data analysis and cloud computing is also an important industry leading the planned development at present. The technology and business of the company are completely consistent with these national industrial policies. At present, the number of chemical plants in China is amazing, and the chemical industry is indeed a potentially huge market for the company's newly developed market. The company's general chemical intelligent optimization solution will help the company's future sustainable business growth in the huge chemical market.

The customer concentration is fairly high, the ratios as the each of top 5 customers sales by the total revenue are 22.38, 15.17, 7.92, 4.39, 3.15%, totaling 53.01%.

2. Financial Analysis

   (1) Basic Financial Data (Table 7.24)

Table 7.24 Basic financial data for HeroopSys from 2017 to 2019 (unit (*yuan* in 10 thousand))

	2019	2018	2017
Total assets	14,180.45	8432.38	6718.82
Net income attributed to shareholders	13,166.89	7344.91	5619.31
Revenue	6228.41	6100.46	4213.47
Net profit	1741.98	2085.60	862.02
Cash flow from operating activities	378.11	1162.86	−34.55

Source Wind, compiled by ZLCA

(2) Primary Financial Data (Table 7.25)

Table 7.25 Primary financial data for HeroopSys from 2017 to 2019

	2019	2018	2017
Gross profit margin (%)	76.78	79.99	80.97
Net profit margin (%)	27.97	34.19	20.46
Debt ratio (%)	7.15	12.90	16.36
Current ratio	13.35	6.78	5.40
Quick ratio	11.38	6.50	4.93
Accounts receivable turnover ratio	1.98	2.06	1.86

Source *Wind*, compiled by ZLCA

(3) Income Structure (Table 7.26)

Table 7.26 The income structure for HeroopSys from 2017 to 2019 (unit *yuan* in ten thousands, %)

Category	2019		2018		2017	
	Amount	Ratio	Amount	Ratio	Amount	Ratio
Software	3861.58	62.00	4803.28	78.74	–	–
Hardware	1657.45	26.61	353.33	5.79	–	–
EPC	419.82	6.74	93.58	1.53	0.00	0.00
Technology development	289.56	4.65	809.21	13.26	0.00	0.00
Service income			40.62	0.67		
Material spare parts			0.43	0.01		
Intelligent optimization control products and services	–	–	–	–	3778.91	89.69
Routine automation projects and others	–	–	–	–	434.56	10.31
Total	6228.41	100.00	6100.45	100.00	4213.47	100.00

Source *Wind*, compiled by ZLCA

7.2.9 Beijing INTOLO Decoration Co., Ltd.

1. Basic Information of the Company

Beijing INTOLO Decoration Co., Ltd. (hereinafter referred to as 'INTOLO') was established in 1996, publicly listed in NEEQ (National Equities Exchange and Quotations) in 2014 (stock code: 831,222.OC). The Market value of the INTOLO is 105 million *yuan* on 31st of December 2019.

The company is an integrated service provider whose core business is residential fine decoration, decoration design and construction services of commercial real estate and other public buildings. On the basis of the original residential hardcover business, the company implements the general contracting mode of hardcover EPC project, and makes use of advanced digital means such as BIM, project information management system and labor management platform to create brand-new residential hardcover integrated products such as residences, apartments and hotels, thus promoting the industrialization and digital transformation and upgrading of the whole decoration industry.

The company is in the architectural decoration industry. The accounts receivable in the building decoration industry are relatively high, and there is a situation that some labor services and projects are subcontracted in the decoration projects of the company. However, due to the influence of macroeconomic and real estate industry policies, the growth rate of the architectural decoration industry slowed down, but the total capacity of the architectural decoration market still rose, and the whole industry developed well.

The customer concentration is fairly high, the ratios as the each of top 5 customers sales by the total revenue are 15.29, 14.40, 6.98, 6.36, 3.87%, totaling 46.90%.

2. Financial Analysis

(1) Basic Financial Data (Table 7.27)

Table 7.27 Basic financial data for INTOLO from 2017 to 2019 (unit (*yuan* in 10 thousand))

	2019	2018	2017
Total assets	117,121.50	89,021.23	74,144.43
Net income attributed to shareholders	23,853.89	19,554.94	17,955.74
Revenue	138,245.43	113,705.08	76,093.76
Net profit	4286.81	1615.53	−3952.73
Cash flow from operating activities	1575.46	1294.46	700.32

Source Wind, compiled by ZLCA

(2) Primary Financial Data (Table 7.28)

Table 7.28 Primary financial data for INTOLO from 2017 to 2019

	2019	2018	2017
Gross profit margin (%)	13.24	10.86	4.64
Net profit margin (%)	3.10	1.42	−5.19
Debt ratio (%)	79.63	78.00	75.78
Current ratio	1.14	1.19	1.21
Quick ratio	0.97	0.97	1.00
Accounts receivable turnover ratio	2.10	2.31	2.11

Source Wind, compiled by ZLCA

(3) Income Structure (Table 7.29)

Table 7.29 The income structure for INTOLO from 2017 to 2019 (unit *yuan* in ten thousands, %)

Category	2019		2018		2017	
	Amount	Ratio	Amount	Ratio	Amount	Ratio
Tooling construction	136,660.10	98.85	111,961.76	98.47	75,045.17	98.62
Design and soft-fitting	848.95	0.61	465.18	0.41	365.07	0.48
Landscape	730.04	0.53	1242.29	1.09	398.36	0.52
Other business	6.34	0.01	35.85	0.03	285.15	0.38
Total	138,245.43	100.00	113,705.08	100.00	76,093.75	100.00

Source Wind, compiled by ZLCA

7.2.10 Ficont Industry (Beijing) Co., Ltd.

1. Basic Information of the Company

Ficont Industry (Beijing) Co., Ltd. (hereinafter referred to as 'Ficont') was established in 2005, publicly listed in NEEQ (National Equities Exchange and Quotations) in 2014 (stock code: 831,222.OC). The Market value of the Ficont is 105 million *yuan* on 31st of December 2019.

The company is a leading provider of high-altitude safe operation equipment and service solutions in China, focusing on the field of high-altitude safe operation, and is committed to providing customers with "safe use, simple operation and professional manufacturing" overall solutions for high-altitude operation. The company's products and services have been applied in more than ten industries such as wind power generation, power grid, communication, thermal power generation, construction, bridges and more than forty countries and regions around the world. At present, the company mainly focuses on the field of wind power generation.

The company's operation is closely related to the development of wind power industry. In order to ensure the proportion of non-fossil energy, promote the transformation and upgrading of energy structure, and promote the sustainable and healthy development of renewable energy represented by wind power, the state has successively issued a number of policies to continue to support the steady development of wind power industry from various aspects such as industry supervision, installation planning and subsidy mechanism. The improvement of consumption situation and the decline of wind abandonment rate will enhance the reinvestment ability of wind power enterprises and enhance the prosperity of the industry; At the same time, the increase in wind power generation will help to improve the economic benefits of wind power enterprises, and hedge the impact of subsidy withdrawal and the reduction of wind power on-grid tariff.

The customer concentration is fairly high, the ratios as the each of top 5 customers sales by the total revenue are 15.51, 15.07, 9.14, 6.56, 6.19%, totaling 52.47%.

2. Financial Analysis

(1) Basic Financial Data (Table 7.30)

Table 7.30 Basic financial data for Ficont industry from 2017 to 2019 (unit (*yuan* in 10 thousand))

	2019	2018	2017
Total assets	101,512.60	82,438.90	67,813.65
Net income attributed to shareholders	77,619.91	66,069.37	58,403.29
Revenue	53,907.70	35,566.75	28,938.10
Net profit	14,177.92	9332.70	7169.22
Cash flow from operating activities	11,000.29	3540.40	4764.07

Source Wind, compiled by ZLCA

(2) Primary Financial Data (Table 7.31)

Table 7.31 Primary financial data for Ficont Industry from 2017 to 2019

	2019	2018	2017
Gross profit margin (%)	55.81	54.89	54.65
Net profit margin (%)	26.30	26.24	24.77
Debt ratio (%)	23.54	19.86	13.88
Current ratio	3.98	4.68	7.06
Quick ratio	3.42	4.23	6.62
Accounts receivable turnover ratio	2.64	2.03	2.05

Source Wind, compiled by ZLCA

(3) Income Structure (Table 7.32)

Table 7.32 The income structure for Ficont industry from 2017 to 2019 (unit *yuan* in ten thousands, %)

Category	2019		2018		2017	
	Amount	Ratio	Amount	Ratio	Amount	Ratio
Special high-altitude safety operation equipment	52,698.75	97.76	34,845.23	97.97	28,305.83	97.82
High altitude safety operation service	997.75	1.85	625.44	1.76	552.64	1.91
Other income	211.20	0.39	96.08	0.27	79.63	0.27
Total	53,907.70	100.00	35,566.75	100.00	28,938.10	100.00

Source Wind, compiled by ZLCA

7.2.11 Beijing Steel Information Network Co., Ltd.

1. Basic Information of the Company

Beijing Steel Information Network Co., Ltd. (hereinafter referred to as 'Steel Information Network') was established in 2009, publicly listed in NEEQ (National Equities Exchange and Quotations) in 2015 (stock code: 831,727.OC). The Market value of the Steel Information Network is 298 million *yuan* on 31st of December 2019.

The company is a global professional WTTI solution provider, which is a national large-scale steel internet enterprise integrating steel electronic trading, online bank payment, steel market information, spot resource search, terminal procurement bidding, steel collection and distribution, and e-commerce services. The company provides customers with standardized commodity spot-based e-commerce trading platform and integrated value-added services based on platform transaction evolution.

The e-commerce industry where the company is involved has a huge transaction scale, and its development is supported by government policies, and its development prospect is very broad. The future market scale of the industry will expand rapidly with the gradual maturity of the market environment and e-commerce platform,. Domestic and foreign enterprises with steel, e-commerce and related industry backgrounds may join the ranks of existing competitors. The increase of competitors will increase the difficulty of maintaining the existing market share and bring potential market and competition risks.

The customer concentration is significantly low, the ratios as the each of top 5 customers sales by the total revenue are 8.05, 4.31, 2.88, 1.74, 1.70%, totaling 18.68%. The company doesn't reveal the specific names of the top 5 customers in 2019.

2. Financial Analysis

 (1) Basic Financial Data (Table 7.33)

Table 7.33 Basic financial data for steel information network from 2017 to 2019 (unit (*yuan* in 10 thousand))

	2019	2018	2017
Total assets	40,964.96	40,475.86	19,914.81
Net income attributed to shareholders	10,933.20	7430.11	2571.26
Revenue	495,213.00	375,550.85	248,147.86
Net profit	3503.10	2165.50	−1617.76
Cash flow from operating activities	−4798.89	3824.10	−1936.30

Source Wind, compiled by ZLCA

(2) Primary Financial Data (Table 7.34)

Table 7.34 Primary financial data for steel information network from 2017 to 2019

	2019	2018	2017
Gross profit margin (%)	3.14	2.73	2.25
Net profit margin (%)	0.71	0.58	−0.65
Debt ratio (%)	73.31	81.64	87.14
Current ratio	1.33	1.22	1.14
Quick ratio	0.77	0.73	0.52
Accounts receivable turnover ratio	111.24	165.28	207.34

Source Wind, compiled by ZLCA

(3) Income Structure (Table 7.35)

Table 7.35 The income structure for steel information network from 2017 to 2019 (unit *yuan* in ten thousands, %)

Category	2019		2018		2017	
	Amount	Ratio	Amount	Ratio	Amount	Ratio
Steel distribution category	494,588.53	99.87	375,217.43	99.91	247,938.16	99.92
Membership fee income	253.19	0.05	179.99	0.05	–	–
Paper media i	140.27	0.03	24.38	0.01	–	–
website construction	121.25	0.02	38.80	0.01	–	–
Promotion fee income	109.76	0.02	–	–	–	–
Promotion fee income	–	–	78.28	0.02	–	–
Other business	–	–	11.97	0.00	–	–
Information service class	–	–	–	–	186.66	0.08
Total	495,213.00	100.00	375,538.88	100.00	248,124.82	100.00

Source Wind, compiled by ZLCA

7.2.12 Beijing Henghe Information and Technology Co., Ltd.

1. Basic Information of the Company

Beijing Henghe Information and Technology Co., Ltd. (hereinafter referred to as 'HENGHE') was established in 2000, publicly listed in NEEQ (National Equities Exchange and Quotations) in 2015 (stock code: 832,145.OC). The Market value of the HENGHE is 263 million *yuan* on 31st of December 2019.

The company is an integrated enterprise engaged in environmental protection technology product development, industrial product application and promotion, and project implementation. The company mainly provides oil and gas recovery online monitoring, online monitoring data management platform, oil leakage detection and other integrated management solutions for petroleum and petrochemical enterprises and environmental protection departments, and provides customers with design, research and development, production, integration, installation and debugging, data analysis and other services.

The environmental protection industry is a typical policy-driven industry. HENGHE has long been committed to the comprehensive prevention and monitoring of atmospheric pollutants VOCs as an environmental protection technology enterprise, and its development is affected by the national environmental protection policy to a certain extent. The domestic policies in the field of energy conservation and environmental protection have been intensified, and the construction of ecological civilization has been defined as a Millennium plan for the sustainable development of the Chinese nation since the 13th Five-Year Plan. The government attaches great importance to environmental issues, and the policies force enterprises to consider environmental protection as an important factor in the development process.

The customer concentration is fairly high, the ratios as the each of top 5 customers sales by the total revenue are 9.44, 9.05, 7.17, 6.75, 6.13%, totaling 38.54%.

2. Financial Analysis

(1) Basic Financial Data (Table 7.36)

Table 7.36 Basic financial data for HENGHE from 2017 to 2019 (unit (*yuan* in 10 thousand))

	2019	2018	2017
Total assets	18,055.92	12,939.29	11,225.94
Net income attributed to shareholders	13,442.94	11,781.94	9685.61
Revenue	12,571.78	10,203.08	6060.31
Net profit	3203.67	2256.95	976.85
Cash flow from operating activities	549.01	1288.55	−224.86

Source *Wind*, compiled by ZLCA

(2) Primary Financial Data (Table 7.37)

Table 7.37 Primary financial data for HENGHE from 2017 to 2019

	2019	2018	2017
Gross profit margin (%)	52.63	40.22	38.67
Net profit margin (%)	25.48	22.12	16.12
Debt ratio (%)	25.55	8.94	13.72
Current ratio	3.85	11.02	7.26
Quick ratio	3.33	8.80	6.11
Accounts receivable turnover ratio	1.90	2.50	1.68

Source *Wind*, compiled by ZLCA

(3) Income Structure (Table 7.38)

Table 7.38 The income structure for HENGHE from 2017 to 2019 (unit *yuan* in ten thousands, %)

Category	2019		2018		2017	
	Amount	Ratio	Amount	Ratio	Amount	Ratio
Online monitoring	8695.87	69.17	5006.91	49.07	2610.54	43.08
Secondary oil and gas recovery system	1981.50	15.76	2872.53	28.15	2690.71	44.40
Liquid level meter system	1332.98	10.60	1952.52	19.14	639.56	10.55
Technical services and others	383.81	3.05	371.12	3.64	119.50	1.97
Leakage detection device of gas station	114.69	0.91	–	–	–	–
Intelligent network online monitoring data management platform	49.70	0.40	–	–	–	–
Testing income	13.23	0.11	–	–	–	–
Total	12,571.78	100.00	10,203.08	100.00	6060.31	100.00

Source Wind, compiled by ZLCA

7.2.13 Beijing Wingain Machinery and Electronics Co., Ltd.

1. Basic Information of the Company

Beijing Wingain Machinery and Electronics Co., Ltd. (hereinafter referred to as 'Wingain Machinery and Electronics') was established in 2011, publicly listed in NEEQ (National Equities Exchange and Quotations) in 2015 (stock code: 832,694.OC). The Market value of the HENGHE is 338 million *yuan* on 31st of December 2019.

The company specializes in providing R&D, design, production and sales of precision metal structures and complete sets of equipment for mobile communication, rail transit, power, oil and gas and other industries. The products include mobile communication base station integrated equipment, rail transit automatic fare collection integrated equipment, large modular control center of energy and power industry, automatic hydraulic power unit and various electrical control cabinets.

The company belongs to computer, communication and other electronic equipment manufacturing industry. The company has not disclosed specific industry information yet.

The customer concentration is significantly high, the ratios as the each of top 5 customers sales by the total revenue are 58.10, 13.17, 9.96, 4.13, 2.18%, totaling 87.54%. The company doesn't reveal the specific names of the top 5 customers in 2019.

2. Financial Analysis

(1) Basic Financial Data (Table 7.39)

Table 7.39 Basic financial data for Wingain machinery and electronics from 2017 to 2019 (unit (*yuan* in 10 thousand))

	2019	2018	2017
Total assets	75,685.10	85,073.50	59,337.02
Net income attributed to shareholders	53,963.34	54,234.62	31,361.41
Revenue	50,778.89	65,321.73	48,332.25
Net profit	2538.71	8404.38	7029.06
Cash flow from operating activities	3511.80	4506.83	1518.15

Source Wind, compiled by ZLCA

(2) Primary Financial Data (Table 7.40)

Table 7.40 Primary financial data for Wingain machinery and electronics from 2017 to 2019

	2019	2018	2017
Gross profit margin (%)	20.86	27.20	33.10
Net profit margin (%)	5.00	12.87	14.54
Debt ratio (%)	28.70	36.25	47.15
Current ratio	2.85	2.34	1.85
Quick ratio	2.42	2.02	1.58
Accounts receivable turnover ratio	2.02	2.62	2.83

Source Wind, compiled by ZLCA

(3) Income Structure (Table 7.41)

Table 7.41 The income structure for Wingain machinery and electronics from 2017 to 2019 (unit *yuan* in ten thousands, %)

Category	2019		2018		2017	
	Amount	Ratio	Amount	Ratio	Amount	Ratio
Mobile communication products	33,405.35	65.79	50,123.46	76.73	38,132.50	78.90
Others	17,373.54	34.21	15,198.27	23.27	10,199.76	21.10
Total	50,778.89	100.00	65,321.73	100.00	48,332.26	100.00

Source Wind, compiled by ZLCA

7.2.14 *Beijing XinZhiHeng Techology Co., Ltd.*

1. Basic Information of the Company

Beijing XinZhiHeng Co., Ltd. (hereinafter referred to as 'XinZhiHeng') was established in 2002, publicly listed in NEEQ (National Equities Exchange and Quotations) in 2015 (stock code: 833,024.OC). The Market value of the XinZhiHeng is 585 million *yuan* on 31st of December 2019.

The company is a high-tech enterprise focusing on Internet of Things technology and focusing on Internet of Things services in the field of intelligent security. The company's main business includes the whole process system integration and service of intelligent security system design, implementation, operation and maintenance in the fields of public security, intelligent transportation, intelligent education, intelligent community and intelligent medical care.

The company is in I65 software and information technology service industry, and the cross-industry competition in recent years has led to the increasing risks faced by enterprises. On the other hand, due to the rapid development of security industry, it may attract more enterprises to enter. The network technology and related products are constantly updated, and new technologies and products are emerging one after another with the increasing demand for personalized products from customers in the video surveillance market.

The customer concentration is fairly low, the ratios as the each of top 4 customers sales by the total revenue are 4.39%, 3.30%, 2.27%, 1.93%, totaling 11.89%.

2. Financial Analysis

   (1) Basic Financial Data (Table 7.42)

Table 7.42 Basic financial data for XinZhiHeng from 2017 to 2019 (unit (*yuan* in 10 thousand))

	2019	2018	2017
Total assets	50,005.55	34,046.63	26,228.14
Net income attributed to shareholders	20,968.27	15,301.37	10,228.22
Revenue	51,992.42	30,292.03	21,519.12
Net profit	5666.90	2201.78	1800.17
Cash flow from operating activities	1395.19	−1594.76	−1463.67

Source: Wind, compiled by ZLCA

(2) Primary Financial Data (Table 7.43)

Table 7.43 Primary financial data for XinZhiHeng from 2017 to 2019

	2019	2018	2017
Gross profit margin (%)	20.39	21.61	20.54
Net profit margin (%)	10.90	7.27	8.37
Debt ratio (%)	58.07	55.06	61.00
Current ratio	1.83	1.80	1.76
Quick ratio	1.14	1.19	1.21
Accounts receivable turnover ratio	5.57	5.14	5.53

Source Wind, compiled by ZLCA

(3) Income Structure (Table 7.44)

Table 7.44 The income structure for XinZhiHeng from 2017 to 2019 (unit *yuan* in ten thousands, %)

Category	2019		2018		2017	
	Amount	Ratio	Amount	Ratio	Amount	Ratio
Installation income	36,807.10	70.79	21,954.39	72.48	17,264.18	80.23
Equipment sales revenue	10,206.22	19.63	7718.47	25.48	3179.75	14.78
Technical service income	4979.09	9.58	619.17	2.04	1075.19	4.99
Total	51,992.41	100.00	30,292.03	100.00	21,519.12	100.00

Source Wind, compiled by ZLCA

7.2.15 HOLLYCRM Co., Ltd.

1. Basic Information of the Company

HOLLYCRM Co., Ltd. (hereinafter referred to as 'HOLLYCRM') was established in 2002, publicly listed in NEEQ (National Equities Exchange and Quotations) in 2015 (stock code: 833,629.OC). The Market value of the HOLLYCRM is 469 million *yuan* on 31st of December 2019.

The company is a senior enterprise specializing in software development and service of call center and customer relationship management, dedicated to providing customers with competitive solutions, and is a leading full set of call center solutions and service providers with independent intellectual property rights in China. The company's main business includes providing software development services for call center systems, call center cloud services, and maintenance services related to the above products and services.

The development of the company depends on the development of software industry. The software industry of call center application software industry is encouraged by the nation, and the nation has given great support in terms of policies, while strategic emerging industries drive the development of call centers. The upgrading of industry and business model has activated the demand for call centers.

The customer concentration is fairly high, the ratios as the each of top 5 customers sales by the total revenue are 45.24, 12.09, 5.35, 3.18, 3.13%, totaling 68.99%.

2. Financial Analysis

(1) Basic Financial Data (Table 7.45)

Table 7.45 Basic financial data for HOLLYCRM from 2017 to 2019 (unit (*yuan* in 10 thousand))

	2019	2018	2017
Total assets	25,190.76	23,497.23	22,040.55
Net income attributed to shareholders	19,328.20	17,591.30	16,380.50
Revenue	24,110.83	18,392.75	18,134.65
Net profit	3257.66	2974.00	3688.57
Cash flow from operating activities	559.45	1567.17	4367.01

Source: *Wind*, compiled by ZLCA

(2) Primary Financial Data (Table 7.46)

Table 7.46 Primary financial data for HOLLYCRM from 2017 to 2019

	2019	2018	2017
Gross profit margin (%)	47.44	55.72	55.58
Net profit margin (%)	13.51	16.17	20.34
Debt ratio (%)	23.27	25.13	25.68
Current ratio	3.56	3.23	3.09
Quick ratio	2.98	2.55	2.71
Accounts receivable turnover ratio	4.35	4.72	4.77

Source Wind, compiled by ZLCA

(3) Income Structure (Table 7.47)

Table 7.47 The income structure for HOLLYCRM from 2017 to 2019 (unit *yuan* in ten thousands, %)

Category	2019		2018		2017	
	Amount	Ratio	Amount	Ratio	Amount	Ratio
Application software development	17,179.29	71.25	12,646.33	68.76	12,028.71	66.33
Cloud services	5705.05	23.66	3905.38	21.23	4199.97	23.16
maintenance service	1030.24	4.27	1239.80	6.74	1214.40	6.70
Commodity sale	196.24	0.81	601.24	3.27	691.57	3.81
Total	24,110.83	100.00	18,392.75	100.00	18,134.65	100.00

Source Wind, compiled by ZLCA

7.2.16 Tengxuan Travel Group Co., Ltd

1. Basic Information of the Company

Tengxuan Travel Group Co., Ltd. (hereinafter referred to as 'Tengxuan Travel') was established in 2001, publicly listed in NEEQ (National Equities Exchange and Quotations) in 2015 (stock code: 430,005.OC). The Market value of the Tengxuan Travel is 773 million *yuan* on 31st of December 2019.

The company is an international travel agency that can provide domestic and international characteristic travel, high-end customized travel, business travel, annual meeting reward travel and other services for individuals, groups and enterprises. The main business includes providing customers with enterprise travel services such as transportation, business travel management, tourism and awards, and providing enterprise customers with value-added business services such as software services and platform services.

The volume of industry is expanding and the industrial system is becoming more and more perfect with the rapid development of China's tourism industry. The number of enterprises operating tourism industry is numerous and increasing due to the low consumption of tourism resources, low barriers to entry, little investment and decent comprehensive benefits. At present, outbound tourism has become a hot spot of industrial investment, with accelerated industry innovation and integration, and increasingly fierce market competition. Tourism business is greatly affected by macro-economy, and the slowdown of economic growth will lead to certain short-term fluctuations in tourism consumption demand on the rising channel.

The customer concentration is significantly low, the ratios as the each of top 5 customers sales by the total revenue are 5.89, 2.93, 2.25, 1.99, 1.91%, totaling 14.97%. The company doesn't reveal the specific names of the top 5 customers in 2019.

2. Financial Analysis

 (1) Basic Financial Data (Table 7.48)

Table 7.48 Basic financial data for Tengxuan travel group from 2017 to 2019 (unit (*yuan* in 10 thousand))

	2019	2018	2017
Total assets	47,495.95	32,431.10	22,577.84
Net income attributed to shareholders	23,197.99	20,941.28	13,062.06
Revenue	200,216.50	190,056.70	155,598.52
Net profit	3670.62	1456.93	3003.32
Cash flow from operating activities	5661.74	615.62	684.30

Source Wind, compiled by ZLCA

(2) Primary Financial Data (Table 7.49)

Table 7.49 Primary financial data for Tengxuan Travel Group from 2017 to 2019

	2019	2018	2017
Gross profit margin (%)	9.13	9.79	8.93
Net profit margin (%)	1.83	0.77	1.93
Debt ratio (%)	49.56	33.44	40.32
Current ratio	1.54	2.10	2.29
Quick ratio	1.54	2.10	2.29
Accounts receivable turnover ratio	22.11	22.36	20.55

Source: *Wind*, compiled by ZLCA

(3) Income Structure (Table 7.50)

Table 7.50 The income structure for Tengxuan Travel Group from 2017 to 2019 (unit *yuan* in ten thousands, %)

Category	2019		2018		2017	
	Amount	Ratio	Amount	Ratio	Amount	Ratio
Free travel ticketing revenue	189,157.69	94.48	171,140.23	90.05	141,559.31	90.98
Travel management	4878.03	2.44	10,062.02	5.29	1915.43	1.23
Ticket agent	2549.68	1.27	1342.19	0.71	1058.46	0.68
Tourism service income	2239.53	1.12	4445.08	2.34	11,047.23	7.10
Platform income	1053.91	0.53	1363.49	0.72	–	–
Software income	232.52	0.12	95.18	0.05	–	–
Member services	–	–	1608.51	0.85	18.09	0.01
Others	105.13	0.05	–	–	–	–
Total	200,216.49	100.00	190,056.70	100.00	155,580.43	100.00

Source Wind, compiled by ZLCA

7.2.17 Guodiankangneng Technology Stock Co., Ltd.

1. Basic Information of the Company

Guodiankangneng Technology Stock Co., Ltd. (hereinafter referred to as 'Guodiankangneng') was established in 1999, publicly listed in NEEQ (National Equities Exchange and Quotations) in 2015 (stock code: 833,966.OC). The Market value of the Guodiankangneng is 1.979 billion *yuan* on 31st of December 2019.

The company is a one-stop supplier of energy-saving and environmental protection services integrating R&D, consultation, construction and operation. It mainly provides industrial energy-saving services (including energy-saving and environmental protection engineering transformation such as waste heat and pressure power generation, flue gas tail gas treatment and sewage treatment) for rare and precious metals smelting and chemical industries in China; Provide energy-saving engineering services and technical services for public buildings, roads and other facilities.

The waste heat power generation business such as metallurgy and chemical industry, construction and green data center business, flue gas tail gas treatment business, and sewage treatment business are energy-saving and environmental protection industries. The national energy-saving and environmental protection policies are intensively introduced and the intensity continues to increase. The energy-saving and environmental protection policies and measures extend from administrative means to legal, administrative and economic means, and the enthusiasm and initiative of the third party to control pollution are fully mobilized. The green energy-saving and environmental protection industry is gradually becoming a new bright spot and a new pillar under the new economic normal. The industrial development is facing an important opportunity period and has great potential.

The customer concentration is significantly high, the ratios as the each of top 5 customers sales by the total revenue are 71.71, 7.77, 6.05, 4.21, 2.58%, totaling 92.32%. The company doesn't reveal the specific names of the top 5 customers in 2019.

2. Financial Analysis

(1) Basic Financial Data (Table 7.51)

Table 7.51 Basic financial data for Guodiankangneng from 2017 to 2019 (unit (*yuan* in 10 thousand))

	2019	2018	2017
Total assets	90,996.81	70,128.12	44,846.51
Net income attributed to shareholders	60,013.63	48,247.43	33,202.34
Revenue	33,441.42	32,515.79	21,881.91
Net profit	10,565.81	10,146.57	8493.77
Cash flow from operating activities	777.18	−4643.41	1179.49

Source Wind, compiled by ZLCA

(2) Primary Financial Data (Table 7.52)

Table 7.52 Primary financial data for Guodiankangneng from 2017 to 2019

	2019	2018	2017
Gross profit margin (%)	48.12	47.48	54.62
Net profit margin (%)	31.60	31.21	38.82
Debt ratio (%)	33.91	31.13	26.04
Current ratio	2.93	2.92	3.38
Quick ratio	1.64	1.40	3.38
Accounts receivable turnover ratio	2.87	3.99	2.48

Source Wind, compiled by ZLCA

(3) Income Structure (Table 7.53)

Table 7.53 The income structure for HTA from 2017 to 2019 (unit *yuan* in ten thousands, %)

Category	2019		2018		2017	
	Amount	Ratio	Amount	Ratio	Amount	Ratio
project contracting	27,442.65	82.06	29,109.76	89.52	18,935.58	86.54
Technology license and service	4291.22	12.83	650.98	2.00	170.28	0.78
contract energy management	1587.85	4.75	2201.88	6.77	2333.11	10.66
Commodity sales	108.65	0.32	271.56	0.84	–	–
Other business	11.04	0.04	281.62	0.87	–	–
Total	33,441.41	100.00	32,515.80	100.00	21,438.97	97.98

Source Wind, compiled by ZLCA

7.2.18 *Beijing Dragon Century Technology, Inc.*

1. Basic Information of the Company

Beijing Dragon Century Technology,Inc. (hereinafter referred to as 'Dragon Century') was established in 2002, publicly listed in NEEQ (National Equities Exchange and Quotations) in 2015 (stock code: 834,037.OC). The Market value of the Dragon Century is 24 million *yuan* on 31st of December 2019.

The company is a professional service provider engaged in large-scale IT infrastructure construction and management. The company provides IT facilities and IT assets life cycle management services for high-speed growth Internet enterprises. That is to say, the company cooperates with customers in depth, from planning and design of IT infrastructure to procurement and deployment, from large-scale distributed field maintenance to professional resource recovery, and provides in-depth services throughout the life cycle.

The company is an information system integration service, and the downstream enterprises are basically large and medium-sized Internet enterprises. The equipment and services of Internet infrastructure are mainly sold to well-known domestic Internet enterprises such as Beijing Qiyi Century Technology Co., Ltd. The company has strengthened its business development in the small and medium-sized Internet industry, and new Internet customers have increased.

The customer concentration is significantly high, the ratios as the each of top 5 customers sales by the total revenue are 69.32, 3.89, 2.88, 2.85, 2.12%, totaling 81.06%. The company doesn't reveal the specific names of the top 5 customers in 2019.

2. Financial Analysis

(1) Basic Financial Data (Table 7.54)

Table 7.54 Basic financial data for dragon century from 2017 to 2019 (unit (*yuan* in 10 thousand))

	2019	2018	2017
Total assets	16,696.60	30,618.11	23,708.86
Net income attributed to shareholders	9309.42	8529.07	8267.93
Revenue	90,733.83	87,537.01	76,191.17
Net profit	3726.01	3003.47	3333.68
Cash flow from operating activities	14,703.35	592.42	1213.46

Source Wind, compiled by ZLCA

(2) Primary Financial Data (Table 7.55)

Table 7.55 Primary financial data for dragon century from 2017 to 2019

	2019	2018	2017
Gross profit margin (%)	7.18	6.44	7.69
Net profit margin (%)	4.11	3.43	4.38
Debt ratio (%)	44.24	72.14	65.13
Current ratio	2.88	1.39	1.53
Quick ratio	2.12	0.60	0.78
Accounts receivable turnover ratio	11.36	11.78	9.77

Source Wind, compiled by ZLCA

(3) Income Structure (Table 7.56)

Table 7.56 The income structure for dragon century from 2017 to 2019 (unit *yuan* in ten thousands, %)

Category	2019		2018		2017	
	Amount	Ratio	Amount	Ratio	Amount	Ratio
Computer hardware sales and service	85,551.00	94.29	83,870.63	95.81	72,658.07	95.36
Service income	5182.83	5.71	3666.38	4.19	3533.11	4.64
Total	90,733.83	100.00	87,537.01	100.00	76,191.18	100.00

Source Wind, compiled by ZLCA

7.2.19 Beijing Truck Information Technology Co., Ltd.

1. Basic Information of the Company

Beijing Truck Information Technology Co., Ltd. (hereinafter referred to as 'Truck') was established in 2008, publicly listed in NEEQ (National Equities Exchange and Quotations) in 2015 (stock code: 834,063.OC). The Market value of the Truck is 711 million *yuan* on 31st of December 2019.

The company is currently the largest truck portal website and the largest interactive service platform for commercial vehicles in China. The company mainly focuses on the commercial vehicle industry and provides brand promotion and promotion services for many commercial vehicle brands at home and abroad; Integrate strategic data guidance, marketing execution monitoring and effect transformation evaluation in the marketing link to provide digital marketing services for commercial vehicle dealers; Provide the end users of commercial vehicles with the services of car selection, car purchase, car use and car maintenance.

The company belongs to information transmission, software and information technology services-Internet and related services-Internet information services. The Internet industry is developing rapidly, and Internet products and technologies are constantly being updated. Internet industry is a high-tech industry supported and developed by the state, and the state gives great support and encouragement in industrial policy.

The customer concentration is fairly low, the ratios as the each of top 5 customers sales by the total revenue are 6.79, 4.81, 4.56, 4.04, 3.49%, totaling 23.69%.

2. Financial Analysis

   (1) Basic Financial Data (Table 7.57)

Table 7.57 Basic financial data for truck from 2017 to 2019 (unit (*yuan* in 10 thousand))

	2019	2018	2017
Total assets	25,535.22	21,927.46	19,518.67
Net income attributed to shareholders	20,182.02	17,193.27	15,706.29
Revenue	15,710.05	13,017.75	10,541.19
Net profit	2988.68	1487.07	184.63
Cash flow from operating activities	2245.06	3146.08	1936.59

Source Wind, compiled by ZLCA

(2) Primary Financial Data (Table 7.58)

Table 7.58 Primary financial data for truck from 2017 to 2019

	2019	2018	2017
Gross profit margin (%)	71.25	63.59	55.83
Net profit margin (%)	19.02	11.42	1.75
Debt ratio (%)	20.96	21.25	19.16
Current ratio	4.61	4.62	5.07
Quick ratio	4.55	4.53	4.98
Accounts receivable turnover ratio	6.21	5.58	3.57

Source Wind, compiled by ZLCA

(3) Income Structure (Table 7.59)

Table 7.59 The income structure for truck from 2017 to 2019 (unit *yuan* in ten thousands, %)

Category	2019		2018		2017	
	Amount	Ratio	Amount	Ratio	Amount	Ratio
Advertising business	14,197.09	90.37	11,173.19	85.83	8151.46	77.33
Electronic commerce business	1512.95	9.63	1844.56	14.17	2389.73	22.67
Total	15,710.04	100.00	13,017.75	100.00	10,541.19	100.00

Source Wind, compiled by ZLCA

7.2.20 Hecom (Beijing) Technology Co., Ltd.

1. Basic Information of the Company

Hecom (Beijing) Technology Co., Ltd. (hereinafter referred to as 'Hecom Technology') was established in 2009, publicly listed in NEEQ (National Equities Exchange and Quotations) in 2015 (stock code: 834,218.OC). The Market value of the Hecom Technology is 1.73 billion *yuan* on 31st of December 2019.

The company is a complete solution provider for enterprise marketing application management. The company's main business is to provide enterprises with mobile marketing management services based on SaaS and cloud model, with mobile marketing management as the industry entry point, mainly covering enterprise customer management, sales collaborative execution, schedule task management, sales process management, data intelligent analysis, etc.

The company belongs to the software and information technology service industry, and its main business is based on SaaS mode. The characteristics and business model of SaaS industry determine that it is difficult for enterprises in the fast-growing period to make profits. The business model of SaaS industry is characterized by the fact that when a product or service is sold, most of its costs and expenses have been confirmed, but its income needs to be confirmed by stages. The difference between the confirmation time points of income and cost expenses leads to the faster income growth and the more obvious financial losses.

The customer concentration is significantly low, the ratios as the each of top 5 customers sales by the total revenue are 2.00, 1.23, 1.21, 1.09, 0.98%, totaling 6.51%.

2. Financial Analysis

(1) Basic Financial Data (Table 7.60)

Table 7.60 Basic financial data for hecom technology from 2017 to 2019 (unit (*yuan* in 10 thousand))

	2019	2018	2017
Total assets	31,190.73	33,858.81	34,002.24
Net income attributed to shareholders	10,933.11	14,175.66	11,603.66
Revenue	12,718.30	10,950.65	9769.58
Net profit	−3143.91	−5550.61	−7252.09
Cash flow from operating activities	−1233.65	−6039.92	−13,083.72

Source Wind, compiled by ZLCA

(2) Primary Financial Data (Table 7.61)

Table 7.61 Primary financial data for hecom technology from 2017 to 2019

	2019	2018	2017
Gross profit margin (%)	60.87	60.58	68.54
Net profit margin (%)	−24.72	−50.69	−74.23
Debt ratio (%)	51.83	44.74	52.71
Current ratio	1.27	1.51	1.31
Quick ratio	1.27	1.48	1.31
Accounts receivable turnover ratio	34.01	0.00	23.82

Source Wind, compiled by ZLCA

(3) Income Structure (Table 7.62)

Table 7.62 The income structure for hecom technology from 2017 to 2019 (unit *yuan* in ten thousands, %)

Category	2019		2018		2017	
	Amount	Ratio	Amount	Ratio	Amount	Ratio
SAAS platform revenue	8859.49	69.66	7890.22	72.05	8325.32	85.22
Software agent revenue	3167.76	24.91	2587.12	23.63	975.40	9.98
Software development revenue	532.46	4.19	156.33	1.43	321.50	3.29
Others	158.58	1.24	316.98	2.89	147.36	1.51
Total	12,718.29	100.00	10,950.65	100.00	9769.58	100.00

Source Wind, compiled by ZLCA

7.2.21 Beijing Chexun Internet Co., Ltd.

1. Basic Information of the Company

Beijing Chexun Internet Co., Ltd. (hereinafter referred to as 'Chexun Internet') was established in 2008, publicly listed in NEEQ (National Equities Exchange and Quotations) in 2015 (stock code: 834,327.OC). The Market Value of the Chexun Internet is 7.70 billion *yuan* on 31st of December 2019.

The company is an Internet car life service platform with car original IP content as the core and covering the entire value chain of car consumption. The company's main business is to provide car information services for Internet users; to provide car manufacturers and sellers with Internet advertising information services; to develop O2O car sales in the field of Internet car e-commerce.

As an industry that provides information services to the automobile industry, one of the leading industries in the national economy, the development of the automobile Internet information service industry is closely related to the development of the automobile industry. At present, it is an important time for new energy vehicles to enter the Chinese market. In recent years, China's car ownership has continued to rise. my country's auto industry is facing a historical moment in which industrial profits shift from the pre-sales market to the after-sales market.

The customer concentration is fairly high, the ratios as the each of top 5 customers sales by the total revenue are 16.40, 11.98, 10.11, 9.35, 9.33%, totaling 57.17%.

2. Financial Analysis

(1) Basic Financial Data (Table 7.63)

Table 7.63 Basic financial data for Chexun internet from 2017 to 2019 (unit (*yuan* in 10 thousands))

	2019	2018	2017
Total assets	25,792.35	22,053.15	18,004.61
Net income attributed to shareholders	19,288.76	16,620.41	13,450.61
Revenue	23,335.61	23,288.64	14,694.97
Net profit	2617.56	3343.12	2207.84
Cash flow from operating activities	−2338.04	1589.71	−852.85

Source Wind, compiled by ZLCA

(2) Primary Financial Data (Table 7.64)

Table 7.64 Primary financial data for Chexun internet from 2017 to 2019

	2019	2018	2017
Gross profit margin (%)	34.89	36.29	44.08
Net profit margin (%)	11.22	14.36	15.02
Debt ratio (%)	24.53	23.60	23.11
Current ratio	4.15	4.09	4.15
Quick ratio	4.15	4.09	4.01
Accounts receivable turnover ratio	1.41	1.66	1.52

Source Wind, compiled by ZLCA

(3) Income Structure (Table 7.65)

Table 7.65 Income structure of Chexun internet from 2017 to 2019 (unit *yuan* in ten thousands, %)

Category	2019		2018		2017	
	Amount	Ratio	Amount	Ratio	Amount	Ratio
Advertising revenue	23,335.61	100.00	23,046.57	98.96	14,587.07	99.27
Other business	–	–	242.07	1.04	–	–
Member and agency service income	–	–	–	–	86.93	0.59
E-commerce revenue	–	–	–	–	20.96	0.14
Total	23,335.61	100.00	23,288.64	100.00	14,694.96	100.00

Source Wind, compiled by ZLCA

7.2.22 *Wishart Media CO., Ltd.*

1. Basic Information of the Company

Wishart Media CO., Ltd. (hereinafter referred to as 'Wishart Media') was established in 2011, publicly listed in NEEQ (National Equities Exchange and Quotations) in 2015 (stock code: 834,476.OC). The Market Value of the Wishart Media is 6.31 billion *yuan* on 31st of December 2019.

The company has a comprehensive service system, has excellent entertainment integrated marketing capabilities, and is known as the industry's "most integrated film and television marketing company." The business scope mainly includes the formulation of overall marketing strategies for movies, TV dramas, and TV programs, social marketing based on social platforms such as Weibo, WeChat, SNS, BBS, and traditional media, offline activities, and integrated marketing.

The company's industry is affected by the policies of the upstream film and television industry. from 2018 to 2019, the country has been strengthening the control of the film and television industry and film and television projects, and related policies have been tightened. The activity of upstream capital and projects has been slowing down compared with the blowout growth of previous years. Businesses of service providers are also affected.

The customer concentration is fairly high, the ratios as the each of top 5 customers sales by the total revenue are 11.04, 7.29, 5.43, 4.68, 4.51%, totaling 32.95%.

2. Financial Analysis

(1) Basic Financial Data (Table 7.66)

Table 7.66 Basic financial data for of Wishart media from 2017 to 2019 (unit (*yuan* in 10 thousands))

	2019	2018	2017
Total assets	18,324.13	13,649.58	11,848.37
Net income attributed to shareholders	13,061.12	11,090.02	10,072.07
Revenue	10,832.94	14,859.60	10,195.26
Net profit	1778.54	2083.12	2264.94
Cash flow from operating activities	4014.55	−2173.27	−191.96

Source Wind, compiled by ZLCA

(2) Primary Financial Data (Table 7.67)

Table 7.67 Primary financial data for Wishart Media from 2017 to 2019

	2019	2018	2017
Gross profit margin (%)	48.08	32.68	42.29
Net profit margin (%)	16.42	14.02	22.22
Debt ratio (%)	28.84	17.19	14.33
Current ratio	2.76	4.66	5.75
Quick ratio	2.28	4.24	4.94
Accounts receivable turnover ratio	3.00	4.69	9.05

Source Wind, compiled by ZLCA

(3) Income Structure (Table 7.68)

Table 7.68 Income structure of Wishart media from 2017 to 2019 (unit *yuan* in ten thousands, %)

Category	2019		2018		2017	
	Amount	Ratio	Amount	Ratio	Amount	Ratio
Movie marketing revenue	6624.7	61.15	7580.86	51.08	7105.47	69.69
Web drama marketing revenue	1338.51	12.36	2246.07	15.13	1137.28	11.15
Variety show marketing income	495.26	4.57	1330.21	8.96	665.97	6.53
Video production revenue	–	–	44.77	0.30	138.24	1.36
Other marketing income	2149.25	19.84	1231.02	8.29	25.72	0.26
Split income	225.21	2.08	2408.23	16.24	–	–
Total	10,832.93	100.00	14,841.16	100.00	10,195.27	100.00

Source Wind, compiled by ZLCA

7.2.23 Beijing Yunchang Game Technology Co., Ltd.

1. Basic Information of the Company

Beijing Yunchang Game Technology Co., Ltd. (hereinafter referred to as 'Yunchang Game') was established in 2009, publicly listed in NEEQ (National Equities Exchange and Quotations) in 2015 (stock code: 834,785.OC). The Market Value of the Yunchang Game is 6.35 billion *yuan* on 31st of December 2019.

The company is a developer and operator focusing on providing leading mobile games and Internet entertainment products. Relying on its unique disruptive innovation concept and strong technical research and development capabilities, the company has launched a series of high-quality mobile games that have been well received by players; up to now the company has launched: 'The Big Pirate', 'Core Warrior', "I am a big man ", 'Tornado Girl' and many other game products.

As my country's emerging cultural innovation industry, the online game industry is one of the key industries supported by the state and will continue to grow in the future. However, with the continuous expansion of the game industry, various national policies have become increasingly strict and standardized. If the country's relevant industry regulatory policies change, it may bring new changes to the overall industry competition, which may adversely affect the company's operating results.

The customer concentration is significantly high, the ratios as the each of top 5 customers sales by the total revenue are 64.36, 22.09, 2.32, 1.91, 1.39%, totaling 92.07%.

2. Financial Analysis

(1) Basic Financial Data (Table 7.69)

Table 7.69 Basic financial data for Yunchang game from 2017 to 2019 (unit (*yuan* in 10 thousands))

	2019	2018	2017
Total assets	31,777.31	27,819.10	21,825.80
Net income attributed to shareholders	27,814.92	22,248.12	20,002.36
Revenue	24,407.08	20,790.96	15,482.83
Net profit	5566.80	2245.75	660.84
Cash flow from operating activities	13,212.85	−7001.91	5090.76

Source Wind, compiled by ZLCA

(2) Primary Financial Data (Table 7.70)

Table 7.70 Primary financial data for Yunchang game from 2017 to 2019

	2019	2018	2017
Gross profit margin (%)	87.00	79.28	56.40
Net profit margin (%)	22.81	10.80	4.27
Debt ratio (%)	12.47	20.03	8.35
Current ratio	6.99	4.21	9.58
Quick ratio	6.99	4.21	9.58
Accounts receivable turnover ratio	2.57	2.57	4.52

Source Wind, compiled by ZLCA

(3) Income Structure (Table 7.71)

Table 7.71 Income structure of Yunchang game from 2017 to 2019 (unit *yuan* in ten thousands, %)

Category	2019		2018		2017	
	Amount	Ratio	Amount	Ratio	Amount	Ratio
Game revenue	24,407.08	100.00	20,790.96	100.00	15,482.83	100.00
Total	24,407.08	100.00	20,790.96	100.00	15,482.83	100.00

Source Wind, compiled by ZLCA

7.2.24 B&G Holding Company

1. Basic Information of the Company

B&G Holding Company(hereinafter referred to as "B&G") was established in 2012, publicly listed in NEEQ (National Equities Exchange and Quotations) in 2015 (stock code: 834,802.OC). The Market Value of the B&G is 16.06 billion *yuan* on 31st of December 2019.

The company is based on the integration of resources from countries around the world, and provides domestic consumers with global high-quality maternal and child products through the Internet. At present, the main business products cover more than 20 categories such as pregnancy clothing, nutrition, supplies, postpartum body shaping clothing, rehabilitation supplies, baby food, daily necessities, clothing, toys, lathes and bedding.

The company's industry is the "I6420 Internet Information Service Industry in I64 Internet and Related Services", and its main products and services are cross-border e-commerce and maternal and child service businesses focusing on maternal and child products. At present, China's maternal and infant product retail industry is in a stage of rapid development. According to data reported by Essence Securities Research Institute, in 2010, the total size of China's 0–12 year old baby market has exceeded one trillion. The data of the past two decades shows that the market has increased by no less than 10% each year and has been increasing year by year, especially in the past three years, it has entered a stage of rapid development. It is estimated that by 2020, the maternal and infant market will reach 4 trillion *yuan*.

The customer concentration is fairly high, the ratios as the each of top 5 customers sales by the total revenue are 16.43, 9.54, 8.05, 4.72, 3.52%, totaling 42.26%. The company doesn't reveal the specific names of the top 5 customers in 2019.

2. Financial Analysis

   (1) Basic Financial Data (Table 7.72)

Table 7.72 Basic financial data for B&G from 2017 to 2019 (unit (*yuan* in 10 thousands))

	2019	2018	2017
Total assets	16,785.66	10,427.79	8272.59
Net income attributed to shareholders	11,864.97	8584.36	5912.38
Revenue	45,082.71	41,043.56	21,049.13
Net profit	2288.04	2594.13	164.15
Cash flow from operating activities	1178.72	1329.71	−5264.71

Source Wind, compiled by ZLCA

(2) Primary Financial Data (Table 7.73)

Table 7.73 Primary financial data for B&G from 2017 to 2019

	2019	2018	2017
Gross profit margin (%)	12.47	11.06	19.65
Net profit margin (%)	5.08	6.32	0.78
Debt ratio (%)	32.91	18.15	28.05
Current ratio	1.71	2.69	1.78
Quick ratio	1.69	2.61	1.16
Accounts receivable turnover ratio	12.13	21.24	12.54

Source Wind, compiled by ZLCA

(3) Income Structure (Table 7.74)

Table 7.74 Income structure of B&G from 2017 to 2019 (unit *yuan* in ten thousands, %)

Category	2019		2018		2017	
	Amount	Ratio	Amount	Ratio	Amount	Ratio
Merchandising	36,520.09	81.01	37,537.59	91.46	18,793.25	89.28
Affiliate service	8556.37	18.98	3416.47	8.32	996.42	4.74
Promotion service	6.25	0.01	89.51	0.22	1259.47	5.98
Total	45,082.71	100.00	41,043.57	100.00	21,049.14	100.00

Source Wind, compiled by ZLCA

7.2.25 *Hly Information Industry Co., Ltd.*

1. Basic Information of the Company

Highly Information Industry Co.,Ltd (hereinafter referred to as 'Highly') was established in 1999, publicly listed in NEEQ (National Equities Exchange and Quotations) in 2015 (stock code: 835,281.OC). The Market Value of the Highly is 52.22 billion *yuan* on 31st of December 2019.

The company's predecessor, Beijing Hanlinhui Technology Development Co., Ltd., was established in 1993 and restructured into a company limited by shares in 1999. It is a holding subsidiary of TCL Group Co., Ltd. The company takes notebook computer distribution as its main business, operating brands including ThinkPad, HP, Toshiba, ASUS, DELL, APPLE, Samsung, Lenovo notebook computers and their digital products and accessories.

Due to the convenient consumer services provided by the Internet, the way consumers consume through the Internet has gradually solidified and stabilized, and the business growth of online channels has tended to be flat; offline channels have been flattened and developed in depth and more concentrated, making it difficult to expand new channels Continuous improvement, diversified channel models are favored by upstream manufacturers. Online and offline integration of sales channels, regional differences, and changes in consumer consumption patterns pose higher challenges to the company's existing channel management. If the company cannot play its own role With its channel advantages and operational management advantages in response to the impact of changes in the external environment, operations will face downward risks.

The customer concentration is fairly high, the ratios as the each of top 5 customers sales by the total revenue are 22.69, 11.86, 4.10, 1.46, 1.40%, totaling 41.51%.

2. Financial Analysis

(1) Basic Financial Data (Table 7.75)

Table 7.75 Basic financial data for Highly from 2017 to 2019 (unit (*yuan* in 10 thousands))

	2019	2018	2017
Total assets	452,050.91	395,792.34	285,030.86
Net income attributed to shareholders	95,677.20	85,776.52	71,881.89
Revenue	2,083,561.75	1,656,699.26	1,523,460.80
Net profit	21,560.43	19,593.90	20,276.62
Cash flow from operating activities	19,457.84	−8026.53	25,354.05

Source Wind, compiled by ZLCA

(2) Primary Financial Data (Table 7.76)

Table 7.76 Primary financial data for Highly from 2017 to 2019

	2019	2018	2017
Gross Profit Margin (%)	3.59	3.96	3.96
Net Profit Margin (%)	1.03	1.18	1.33
Debt Ratio (%)	77.81	77.59	73.56
Current Ratio	1.29	1.29	1.35
Quick Ratio	0.53	0.39	0.60
Accounts Receivable Turnover Ratio	27.60	29.08	27.59

Source Wind, compiled by ZLCA

(3) Income Structure (Table 7.77)

Table 7.77 Income structure of Highly from 2017 to 2019 (unit *yuan* in ten thousands, %)

Category	2019		2018		2017	
	Amount	Ratio	Amount	Ratio	Amount	Ratio
Distribution	1,223,320.16	58.71	1,164,606.71	70.30	1,041,787.13	68.54
Supply chain and e-commerce services	492,312.37	23.63	292,344.87	17.65	322,483.65	21.22
Retail	207,886.97	9.98	151,801.98	9.16	115,670.44	7.61
Enterprise	156,311.09	7.50	44,904.51	2.71	40,115.4	2.63
Other business	3731.15	0.18	3041.19	0.18	–	–
Total	2,083,561.74	100.00	1,656,699.26	100.00	1,520,056.62	100.00

Source: *Wind*, compiled by ZLCA

7.2.26 Beijing Holyzone Technology Corporation

1. Basic Information of the Company

Beijing Holyzone Technology Corporation (hereinafter referred to as 'Holyzone') was established in 2001, publicly listed in NEEQ (National Equities Exchange and Quotations) in 2006 (stock code: 430,005.OC). The Market Value of the Holyzone is 5.037 billion *yuan* on 31st of December 2019.

The company is committed to providing customers with professional IT solutions, mainly including application delivery, network security, core network services and other fields, and providing them with related technical support and value-added services.

The company's industry is the information technology industry. The development of the information technology industry presents the characteristics of rapid technological upgrading and product upgrading. Enterprises need to have the ability to predict industry development trends, adjust the industry direction in time, and continuously lead or maintain consumer demand for products in order to have long-term development.

The customer concentration is fairly low, the ratios as the each of top 5 customers sales by the total revenue are 6.16, 5.84, 4.82,,4.42, 3.13%, totaling 24.37%.

2. Financial Analysis

(1) Basic Financial Data (Table 7.78)

Table 7.78 Basic financial data for Holyzone from 2017 to 2019 (unit (*yuan* in 10 thousands))

	2019	2018	2017
Total assets	17,001.39	13,890.14	12,357.57
Net income attributed to shareholders	14,950.68	12,670.88	11,442.67
Revenue	49,663.17	36,773.33	30,344.59
Net profit	2682.53	2109.58	2158.94
Cash flow from operating activities	2001.27	2314.46	313.59

Source Wind, compiled by ZLCA

(2) Primary Financial Data (Table 7.79)

Table 7.79 Primary financial data for Holyzone from 2017 to 2019

	2019	2018	2017
Gross profit margin (%)	15.79	18.05	19.66
Net profit margin (%)	5.40	5.74	7.11
Debt ratio (%)	12.06	8.78	7.40
Current ratio	8.15	11.18	13.35
Quick ratio	5.11	5.97	7.16
Accounts receivable turnover ratio	12.39	11.78	12.01

Source Wind, compiled by ZLCA

(3) Income Structure (Table 7.80)

Table 7.80 Income structure of Holyzone from 2017 to 2019 (unit *yuan* in ten thousands, %)

Category	2019		2018		2017	
	Amount	Ratio	Amount	Ratio	Amount	Ratio
Application delivery products	31,657.52	63.74	24,196.91	65.80	20,373.69	67.14
Service income	5750.03	11.58	5445.29	14.81	4070.88	13.42
Wireless network products	5324.67	10.72	–	–	–	–
Other product revenue	3556.99	7.16	4001.42	10.88	2415.89	7.96
Cyber Security Products	3287.42	6.62	3129.71	8.51	3481.86	11.47
Other business	86.54	0.18	–	–	2.28	0.01
Total	49,663.17	100.00	36,773.33	100.00	30,344.6	100.00

Source Wind, compiled by ZLCA

7.2.27 *Beijing Suntray Technology Corporation*

1. Basic Information of the Company

Beijing Suntray Technology Corporation (hereinafter referred to as 'Suntray') was established in 2007, publicly listed in NEEQ (National Equities Exchange and Quotations) in 2016 (stock code: 835,363.OC).

The company is a software and IT service provider based in China, specializing in application software development and maintenance, software quality testing/guarantee, system architecture, business process outsourcing services, software outsourcing, IT technology services, Internet applications, business intelligence, Personnel assignment, etc., the company always provides the best quality service at the most competitive price in the world.

The software and information technology service industry has the characteristics of rapid technological progress, short product life cycles, and frequent upgrades. Enterprises must accurately judge industry technology and application development trends to position their own technology and product development directions, continue to innovate, and continuously introduce new products. Products to meet market demand. When the industry continues to rapidly upgrade and multiple technologies develop simultaneously, precise positioning of the technology development direction puts forward higher requirements for enterprises. This not only requires companies to be forward-looking in market judgment, but also requires strong support in the construction of technical reserves and talent team, and also requires corresponding financial strength to support the realization of technological innovation.

The customer concentration is fairly high, the ratios as the each of top 5 customers sales by the total revenue are 16.95, 7.05, 6.52, 3.72, 2.49%, totaling 36.73%.

2. Financial Analysis

(1) Basic Financial Data (Table 7.81)

Table 7.81 Basic financial data for suntray from 2017 to 2019 (unit (*yuan* in 10 thousands))

	2019	2018	2017
Total assets	9967.95	7794.13	6842.20
Net income attributed to shareholders	5760.07	5320.47	4359.55
Revenue	15,769.05	14,010.76	10,773.09
Net profit	2244.55	1604.22	1208.73
Cash flow from operating activities	1624.96	738.64	−387.47

Source Wind, compiled by ZLCA

(2) Primary Financial Data (Table 7.82)

Table 7.82 Primary financial data for suntray from 2017 to 2019

	2019	2018	2017
Gross profit margin (%)	36.79	37.28	37.79
Net profit margin (%)	14.23	11.45	11.22
Debt ratio (%)	42.09	31.58	36.15
Current ratio	2.36	3.24	2.75
Quick ratio	2.09	3.07	2.63
Accounts receivable turnover ratio	3.17	3.73	4.40

Source Wind, compiled by ZLCA

(3) Income Structure (Table 7.83)

Table 7.83 Income structure of suntray from 2017 to 2019 (unit *yuan* in ten thousands, %)

Category	2019		2018		2017	
	Amount	Ratio	Amount	Ratio	Amount	Ratio
IT talent outsourcing service	15,179.31	96.26	13,252.17	94.59	9571.79	88.85
Project R&D Service	554.94	3.52	707.89	5.05	1616.54	15.01
Business Process Outsourcing Service	34.8	0.22	50.69	0.36	108.04	1.00
Inter-segment elimination	−523.27	−4.86	–	–	–	–
Total	15,769.05	100.00	14,010.75	100.00	10,773.1	100.00

Source Wind, compiled by ZLCA

7.2.28 Beijing Agrichina Pharmaceutical Co., Ltd.

1. Basic Information of the Company

Beijing Agrichina Pharmaceutical Co., Ltd. (hereinafter referred to as 'Agrichina Pharmaceutical') was established in 1986, publicly listed in NEEQ (National Equities Exchange and Quotations) in 2016 (stock code: 835,911.OC). The Market Value of the Agrichina Pharmaceutical is 4.75 billion *yuan* on 31st of December 2019.

The company is a national high-tech enterprise mainly engaged in the production, sales and technical service of veterinary preparations. As a specialized company in the veterinary medicine manufacturing industry, Zhongnong Huawei is committed to the research and development, production, sales and technical services of veterinary medicines. The products produced are mainly used for the prevention and treatment of livestock and poultry diseases. The company has a number of invention patented technologies such as long-acting injections containing ethyl cellulose and fatty acid esters, preparations containing antimicrobial drugs with silicone oil as the medium, and can transform core technologies into superior products.

With the rapid development of China's animal husbandry, the output value of animal husbandry continues to increase, and the company's veterinary drug market sales will continue to grow. Changes in market demand have prompted changes in veterinary drug companies. Veterinary drug companies must not only meet customer needs from a single disease treatment to a systematic preventive health care, but also continuously strengthen the breadth and depth of technical services and continuously develop new varieties of veterinary drugs to cope with complexities. Disease.

The customer concentration is fairly high, the ratios as the each of top 5 customers sales by the total revenue are 39.25, 3.20, 2.37, 2.20, 1.87%, totaling 48.89%.

2. Financial Analysis

(1) Basic Financial Data (Table 7.84)

Table 7.84 Basic financial data for Agrichina pharmaceutical from 2017 to 2019 (unit (*yuan* in 10 thousands))

	2019	2018	2017
Total assets	29,422.75	20,057.6	15,172.93
Net income attributed to shareholders	14,488.58	12,533.14	10,302.68
Revenue	21,508.95	15,565.61	12,337.84
Net profit	2099.56	1271.88	−450.14
Cash flow from operating activities	5422.03	441.82	−1929.05

Source Wind, compiled by ZLCA

(2) Primary Financial Data (Table 7.85)

Table 7.85 Primary financial data for Agrichina pharmaceutical from 2017 to 2019

	2019	2018	2017
Gross profit margin (%)	25.29	28.86	24.09
Net profit margin (%)	9.76	8.17	-3.65
Debt ratio (%)	47.16	36.51	31.16
Current ratio	1.58	2.36	2.88
Quick ratio	1.27	1.99	1.77
Accounts receivable turnover ratio	3.10	2.42	3.44

Source Wind, compiled by ZLCA

(3) Income Structure (Table 7.86)

Table 7.86 Income structure of Agrichina pharmaceutical from 2017 to 2019 (unit *yuan* in ten thousands, %)

Category	2019		2018		2017	
	Amount	Ratio	Amount	Ratio	Amount	Ratio
Powder	11,898.93	57.50	9403.04	62.59	7519.53	62.67
Tablet	2842.68	13.74	2940.30	19.57	1806.07	15.05
Feed	1973.32	9.54	1343.79	8.94	2071.01	17.26
Injection	1665.20	8.05	1336.95	8.90	276.31	2.30
Vaccine	1620.44	7.83	–	–	–	–
Original powder	691.66	3.34	–	–	326.24	2.72
Total	20,692.23	100.00	15,024.08	100.00	11,999.16	100

Source Wind, compiled by ZLCA

7.2.29 Beijing Huiheng Environmental Protection Engineering Co., Ltd.

1. Basic Information of the Company

Beijing Huiheng Environmental Protection Engineering Co., Ltd. (hereinafter referred to as 'Huiheng Environmental Protection') was established in 2003, publicly listed in NEEQ (National Equities Exchange and Quotations) in 2016 (stock code: 836,348.OC). The Market Value of the Huiheng Environmental Protection is 3.30 billion *yuan* on 31st of December 2019.

The company is a high-tech enterprise specializing in sewage and wastewater treatment projects, water supply treatment projects and flue gas desulfurization and dust removal projects. As a provider of overall solutions for sewage treatment projects, the company's products and services include sewage treatment projects (exploded in EPC mode, mainly through bidding or business negotiation to obtain projects), separate sales of environmentally friendly products, custody operation services, and environmental protection projects related design service. Part or all of the links involving sewage treatment project design, equipment sales, project construction, equipment installation, commissioning and operation, technical support, and trusteeship operations. The company's business model includes R&D, procurement, sales and profitability.

The environmental protection industry in which the company is located has developed rapidly in recent years. However, the environmental protection industry is difficult to finance, and projects generally adopt the PPP model, which is characterized by heavy assets and long-term cycles. The approval cycle of special loans for PPP projects is long. Before financing is in place, all project funds are advanced by the company. Cash flow has become a key factor restricting the sustainable operation of environmental protection companies. Environmental protection companies that are more sensitive to the external financing environment will be more affected.

The customer concentration is significantly high, the ratios as the each of top 5 customers sales by the total revenue are 28.25, 25.61, 10.89, 5.30, 3.61%, totaling 73.66%.

2. Financial Analysis

(1) Basic Financial Data (Table 7.87)

Table 7.87 Basic financial data for Huiheng environmental protection from 2017 to 2019 (unit (*yuan* in 10 thousands))

	2019	2018	2017
Total assets	41,614.48	40,737.09	35,688.17
Net income attributed to shareholders	22,841.54	21,789.03	19,439.46
Revenue	9255.23	13,962.38	19,756.67
Net profit	1572.51	2349.57	2187.61
Cash flow from operating activities	−3443.37	−4567.78	−2116.76

Source Wind, compiled by ZLCA

(2) Primary Financial Data (Table 7.88)

Table 7.88 Primary financial data for Huiheng environmental protection from 2017 to 2019

	2019	2018	2017
Gross profit margin (%)	37.45	32.65	21.54
Net profit margin (%)	16.99	16.83	11.07
Debt ratio (%)	45.11	46.51	45.53
Current ratio	2.05	2.00	2.17
Quick ratio	1.15	0.94	0.93
Accounts Receivable Turnover Ratio	1.16	2.78	7.96

Source Wind, compiled by ZLCA

(3) Income Structure (Table 7.89)

Table 7.89 Income structure of Huiheng environmental protection from 2017 to 2019 (unit *yuan* in ten thousands, %)

Category	2019		2018		2017	
	Amount	Ratio	Amount	Ratio	Amount	Ratio
Green facilities	4323.65	46.72	11,548.96	82.71	18,833.28	95.36
Environmental engineering	4034.50	43.59	1273.58	9.12	646.88	3.28
Design technical service	589.50	6.37	997.50	7.14	36.66	0.19
Operation service	297.94	3.22	137.25	0.98	231.95	1.17
Other business	9.64	0.10	5.09	0.04	–	–
Total	9255.23	100.00	13,962.38	100.00	19,748.77	100.00

Source Wind, compiled by ZLCA

7.2.30 Beijing Huachengzhiyun Soft Co., Ltd.

1. Basic Information of the Company

Beijing Huachengzhiyun Soft Co., Ltd (hereinafter referred to as 'Huachengzhiyun') was established in 2011, publicly listed in NEEQ (National Equities Exchange and Quotations) in 2016 (stock code: 836,464.OC). The Market Value of the Huachengzhiyun is 0.60 billion *yuan* on 31st of December 2019.

The company is an excellent domestic software and application integration service provider. The company has been committed to the research, development and application of professional technical services in the field of public security and intelligent transportation in the security industry, and provides application software development, technical support and services to large-scale safe city projects and intelligent transportation projects. At present, the company's main business is video surveillance platform software development and sales, professional solutions and technical services in the security field, and operation and maintenance services.

The company's software and application integration technology are updated quickly, and market demand changes quickly. With the development of information technology, market competition has intensified, and users in various industries have increasingly higher requirements for technology. Companies in the industry attach great importance to technological innovation and market development.

The customer concentration is fairly high, the ratios as the each of top 5 customers sales by the total revenue are 19.3, 11.68, 7.74, 6.9, 5.71%, totaling 51.33%.

2. Financial Analysis

(1) Basic Financial Data (Table 7.90)

Table 7.90 Basic financial data for Huachengzhiyun from 2017 to 2019 (unit (*yuan* in 10 thousands))

	2019	2018	2017
Total assets	17,384.07	13,969.00	11,746.24
Net income attributed to shareholders	11,919.71	9381.46	8223.04
Revenue	11,764.64	9907.27	8459.43
Net profit	2507.81	1134.81	1153.02
Cash flow from operating activities	−1832.56	−627.32	−2662.44

Source Wind, compiled by ZLCA

(2) Primary Financial Data (Table 7.91)

Table 7.91 Primary financial data for Huachengzhiyun from 2017 to 2019

	2019	2018	2017
Gross profit margin (%)	40.66	35.29	27.96
Net profit margin (%)	21.32	11.45	13.63
Debt ratio (%)	31.80	33.02	22.98
Current ratio	2.26	1.90	3.17
Quick ratio	2.03	1.89	3.11
Accounts receivable turnover ratio	1.52	1.64	2.23

Source Wind, compiled by ZLCA

(3) Income Structure (Table 7.92)

Table 7.92 Income structure of Huachengzhiyun from 2017 to 2019 (unit *yuan* in ten thousands, %)

Category	2019		2018		2017	
	Amount	Ratio	Amount	Ratio	Amount	Ratio
Software development and sales	10,059.43	85.51	5031.89	50.79	5112.81	60.44
Technical Services	1422.79	12.09	4581.98	46.25	2405.00	28.43
Operation and maintenance	282.42	2.40	293.40	2.96	941.62	11.13
total	11,764.64	100.00	9907.27	100.00	8459.43	100.00

Source Wind, compiled by ZLCA

7.2.31 *China Bidding Co., Ltd.*

1. Basic Information of the Company

CHINA BIDDING CO., LTD. (hereinafter referred to as 'CHINA BIDDING') was established in 2001, publicly listed in NEEQ (National Equities Exchange and Quotations) in 2006 (stock code: 430,005.OC). The Market Value of the CHINA BIDDING is 5.037 billion *yuan* on 31st of December 2019.

The company is a high-tech enterprise that uses the Internet to engage in bidding transaction data service platform operations. The main business is to rely on the research and development of the electronic bidding procurement technology service platform to provide data mining and processing services for the entire service chain of the bidding industry for all bidding parties such as tenderers, bidders, and bidders. Practitioners such as bidding project managers provide pre-examination training, registration guidance and continuing education services for the qualification and level examination of the tenderer, and provide services such as credit evaluation, brand promotion, corporate image, and product display for bidding companies, and for bidders and bidders The agency provides R&D and operation services for electronic bidding procurement transactions and management platforms.

The company's main business industry information services and electronic bidding are supported by policies and are in line with industry development trends. The electronic bidding business market is growing rapidly, and the company has a foundation for sustainable development. The market prospects are good, and the competition in the industry is fierce.

The customer concentration is significantly low, the ratios as the each of top 5 customers sales by the total revenue are 2.43, 0.62, 1.59, 0.96, 0.85%, totaling 7.45%. The company doesn't reveal the specific names of the top 5 customers in 2019.

2. Financial Analysis

    (1) Basic Financial Data (Table 7.93)

Table 7.93 Basic financial data for China biddinG from 2017 to 2019 (unit (*yuan* in 10 thousands))

	2019	2018	2017
Total assets	21,622.18	18,716.68	17,296.96
Net income attributed to shareholders	17,499.57	15,707.46	14,296.3
Revenue	14,804.87	13,724.39	12,045.66
Net profit	1943.287	1899.614	2705.242
Cash flow from operating activities	3665.003	1133.147	1651.065

Source Wind, compiled by ZLCA

(2) Primary Financial Data (Table 7.94)

Table 7.94 Primary financial data for China bidding from 2017 to 2019

	2019	2018	2017
Gross profit margin (%)	27.82	34.79	44.98
Net profit margin (%)	13.13	13.84	22.46
Debt ratio (%)	16.94	13.61	14.67
Current ratio	4.97	6.29	5.83
Quick ratio	4.97	6.29	5.82
Accounts receivable turnover ratio	48.82	105.17	445.6

Source Wind, compiled by ZLCA

(3) Income Structure (Table 7.95)

Table 7.95 Income structure of China bidding from 2017 to 2019 (unit *yuan* in ten thousands, %)

Category	2019		2018		2017	
	Amount	Ratio	Amount	Ratio	Amount	Ratio
Bidding information platform	7041.45	47.56	7094.33	51.69	8814.45	73.18
Electronic bidding service	5764.24	38.93	3972.78	28.95	390.67	3.24
Consultation service	1903.42	12.86	2638.24	19.22	2432.00	20.19
Other business	95.76	0.65	19.05	0.14	–	–
Industry training	–	–	–	–	408.54	3.39
Total	14,804.87	100.00	13,724.40	100.00	12,045.66	100.00

Source Wind, compiled by ZLCA

7.2.32 Beijing Sunway Technology Co., Ltd.

1. Basic Information of the Company

Beijing Sunway Technology Co.,Ltd. (hereinafter referred to as 'Sunway Technology') was established in 1989, publicly listed in NEEQ (National Equities Exchange and Quotations) in 2016 (stock code: 836,870.OC). The Market Value of the Sunway Technology is 2.35 billion *yuan* on 31st of December 2019.

The company is a Beijing high-tech enterprise, Zhongguancun high-tech enterprise and "double-soft certification" enterprise specializing in software development, software product sales and geographic information system engineering construction in the field of geographic information (including 3S technology). The company specializes in the field of geographic information (including 3S technology) software development, software product sales, technical services and geographic information system engineering construction. It is a leading domestic developer of GIS data production, database construction, and update management software and provides comprehensive geographic information industry Solution provider.

The industry the company belongs to is the software and information technology service industry, which has good market development prospects. With the development of new technologies such as Internet cloud storage, cloud computing, and big data, it means that surveying, mapping, geographic information and other industries are working together to create new technologies, new models, and new products. With the deepening of the "Internet of Things + " in the surveying and mapping geographic information industry and the development and popularization of 5G technology, a profound change will occur in the surveying and mapping geographic information.

The customer concentration is fairly high, the ratios as the each of top 5 customers sales by the total revenue are 18.45, 4.41, 4.31, 3.42, 2.48%, totaling 33.07%.

2. Financial Analysis

(1) Basic Financial Data (Table 7.96)

Table 7.96 Basic financial data for sunway technology from 2017 to 2019 (unit (*yuan* in 10 thousands))

	2019	2018	2017
Total assets	12,988.58	11,806.23	8902.75
Net income attributed to shareholders	11,864.96	10,430.81	7963.34
Revenue	9397.19	8717.38	6433.32
Net profit	2609.15	2009.96	1259.69
Cash flow from operating activities	2559.02	3090.53	302.48

Source Wind, compiled by ZLCA

(2) Primary Financial Data (Table 7.97)

Table 7.97 Primary financial data for sunway technology from 2017 to 2019

	2019	2018	2017
Gross profit margin (%)	65.30	56.71	41.91
Net profit margin (%)	27.77	23.06	19.58
Debt ratio (%)	8.65	11.65	10.55
Current ratio	8.55	6.02	5.74
Quick ratio	8.29	5.77	5.35
Accounts receivable turnover ratio	4.21	3.79	3.62

Source Wind, compiled by ZLCA

(3) Income Structure (Table 7.98)

Table 7.98 Income structure of sunway technology from 2017 to 2019 (unit *yuan* in ten thousands, %)

Category	2019		2018		2017	
	Amount	Ratio	Amount	Ratio	Amount	Ratio
Technical services	4648.14	49.46	4618.00	52.97	–	–
Software revenue	4565.55	48.58	3404.19	39.05	3413.23	53.06
Technology development income	183.50	1.95	671.00	7.70	2902.06	45.11
Other business	0.00	0.00	24.18	0.28	118.03	1.83
Total	9397.19	100.00	8717.38	100.00	6433.32	100.00

Source Wind, compiled by ZLCA

7.2.33 Beijing Headspring Technology Co., Ltd.

1. Basic Information of the Company

Beijing Headspring Technology Co., Ltd. (hereinafter referred to as 'Headspring Technology') was established in 2003, publicly listed in NEEQ (National Equities Exchange and Quotations) in 2016 (stock code: 836,885.OC). The Market Value of the Headspring Technology is 1.17 billion *yuan* on 31st of December 2019.

The company is mainly engaged in transportation informatization, transportation-specific product development, system integration, commercial operation and other comprehensive transportation informatization services. It is an integrated solution provider and service provider in the field of intelligent transportation. The company mainly provides integrated information solutions for highway traffic emergency, transportation, planning, maintenance, road administration and integrated management platform for administrative agencies and enterprises and institutions.

The intelligent transportation industry where the company is located is an industry that the state focuses on supporting and encouraging development. The investment scale of China's intelligent transportation system construction is still small. With the rapid growth of new transportation infrastructure, there is also a huge amount of maintenance and upgrading of existing intelligent transportation systems. Demand space. At present, the overall intelligent transportation market is relatively fragmented, the industry market concentration is low, and the regionality is strong. With the development and application of new technologies, integrated transportation intelligent collaborative services, safe operation, and cross-border integration will be the general development trend.

The customer concentration is fairly high, the ratios as the each of top 5 customers sales by the total revenue are 9.46, 8.75, 8.50, 7.99, 6.43% totaling 41.13%.

2. Financial Analysis

(1) Basic Financial Data (Table 7.99)

Table 7.99 Basic financial data for headspring technology from 2017 to 2019 (unit (*yuan* in 10 thousands))

	2019	2018	2017
Total assets	24,142.58	19,478.07	16,523.79
Net income attributed to shareholders	18,527.12	15,823.30	13,656.91
Revenue	14,212.25	12,357.49	11,853.88
Net profit	2439.43	2136.49	1383.39
Cash flow from operating activities	976.12	−1232.91	−2222.03

Source Wind, compiled by ZLCA

(2) Primary Financial Data (Table 7.100)

Table 7.100 Primary financial data for headspring technology from 2017 to 2019

	2019	2018	2017
Gross profit margin (%)	48.92	45.91	36.84
Net profit margin (%)	17.16	17.29	11.67
Debt ratio (%)	21.36	15.50	15.49
Current ratio	4.40	6.06	6.20
Quick ratio	3.95	5.39	5.76
Accounts receivable turnover ratio	1.70	2.12	2.98

Source Wind, compiled by ZLCA

(3) Income Structure (Table 7.101)

Table 7.101 Income structure of headspring technology from 2017 to 2019 (unit *yuan* in ten thousands, %)

Category	2019		2018		2017	
	Amount	Ratio	Amount	Ratio	Amount	Ratio
Technology development service	8608.94	60.57	6652.80	53.84	7841.75	66.15
Product sales	5603.31	39.43	5704.70	46.16	4012.13	33.85
Total	14,212.25	100.00	12,357.49	100.00	11,853.88	100.00

Source Wind, compiled by ZLCA

7.2.34 China Electronic Cowan Science and Technology Co., Ltd.

1. Basic Information of the Company

China Electronic Cowan Science and Technology Co., Ltd. (hereinafter referred to as 'China Electronic Cowan Science and Technology') was established in 2010, publicly listed in NEEQ (National Equities Exchange and Quotations) in 2016 (stock code: 837,840.OC). The Market Value of the China Electronic Cowan Science and Technology is 9.50 billion *yuan* on 31st of December 2019.

The company is a leading domestic security and emergency management operation service provider. The company's main business includes security and emergency management consulting and design, system integration, big data cloud services, operation services and the development and sales of related software and hardware products.

The company's security emergency and power Internet of Things business segments are in a favorable market environment. The power Internet of Things industry vigorously promotes the construction of power Internet of Things. In 2020, the State Grid clearly proposes to comprehensively build an internationally leading energy Internet enterprise with Chinese characteristics and make every effort to promote High-quality development of power Internet of Things. The company has always paid attention to R&D investment, and has been improving the transformation of scientific and technological innovation results and intellectual property protection. In the future, it will further enhance its vitality and market competitiveness through management innovation and system optimization.

The customer concentration is significantly high, the ratios as the each of top 5 customers sales by the total revenue are 20.15, 15.32, 14.54, 10.93, 9.91%, totaling 70.85%.

2. Financial Analysis

(1) Basic Financial Data (Table 7.102)

Table 7.102 Basic financial data for China electronic cowan science and technology from 2017 to 2019 (unit (*yuan* in 10 thousands))

	2019	2018	2017
Total assets	26,937.12	25,656.71	25,430.19
Net income attributed to shareholders	21,420.76	20,596.44	21,383.50
Revenue	7938.03	7348.60	8580.66
Net profit	1975.44	1951.13	2848.82
Cash flow from operating activities	−1767.78	1923.96	209.84

Source Wind, compiled by ZLCA

(2) Primary Financial Data (Table 7.103)

Table 7.103 Primary financial data for China electronic cowan science and technology from 2017 to 2019

	2019	2018	2017
Gross profit margin (%)	66.26	53.79	59.14
Net profit margin (%)	24.89	26.55	33.20
Debt ratio (%)	20.65	19.76	15.32
Current ratio	4.46	4.83	6.38
Quick ratio	4.19	4.49	6.26
Accounts receivable turnover ratio	0.51	0.52	0.72

Source Wind, compiled by ZLCA

(3) Income Structure (Table 7.104)

Table 7.104 Income structure of China electronic cowan science and technology from 2017 to 2019 (unit *yuan* in ten thousands, %)

Category	2019		2018		2017	
	Amount	Ratio	Amount	Ratio	Amount	Ratio
Hardware products	1841.84	23.20	3342.00	45.48	4948.09	57.67
Software products	4802.31	60.50	3673.59	49.99	3330.34	38.81
Technical services	1293.88	16.30	333.02	4.53	201.82	2.35
Other	0.00	0.00	0.00	0.00	100.41	1.17
Total	7938.03	100	7348.60	100	8580.65	100

Source Wind, compiled by ZLCA

7.2.35 Beijing Orient Tonghua Technology Co., Ltd.

1. Basic Information of the Company

Beijing Orient Tonghua Technology Co., Ltd. (hereinafter referred to as 'Orient Tonghua') was established in 2000, publicly listed in NEEQ (National Equities Exchange and Quotations) in 2016 (stock code: 837,899.OC). The Market Value of the Orient Tonghua is 7.72 billion *yuan* on 31st of December 2019.

The company is a high-tech enterprise that provides comprehensive environmental governance services and products. The company integrates the research and development, sales, engineering and operation of domestic sewage, domestic and kitchen waste treatment equipment, and provides customers with environmental protection and governance Overall solution. The company's main customers are government departments and enterprises all over the country.

The environmental protection industry where the company is located lags behind economic development. The environmental protection industry is still in a stage that requires vigorous development, and industry competition is moderate. In the future, as the country actively encourages and supports the promotion of policies and a large influx of industrial capital, the industry as a whole will be fiercely competitive, and the overall gross profit rate of enterprises in the industry will inevitably decrease. At the same time, the company's industry has typical technology-promoting characteristics. With the improvement of national environmental protection standards, higher requirements are put forward for environmental pollution control technology.

The customer concentration is significantly high, the ratios as the each of top 5 customers sales by the total revenue are 20.00%, 17.00%, 14.00%, 11.00%, 8.00%, totaling 70.00%.

2. Financial Analysis

(1) Basic Financial Data (Table 7.105)

Table 7.105 Basic financial data for orient Tonghua from 2017 to 2019 (unit (*yuan* in 10 thousands))

	2019	2018	2017
Total assets	59,050.91	50,344.47	41,104.81
Net income attributed to shareholders	31,557.02	22,788.94	20,864.84
Revenue	19,252.87	17,266.15	16,307.46
Net profit	5973.19	5099.75	3990.79
Cash flow from operating activities	−718.70	−775.31	−488.74

Source Wind, compiled by ZLCA

(2) Primary Financial Data (Table 7.106)

Table 7.106 Primary financial data for orient Tonghua from 2017 to 2019

	2019	2018	2017
Gross profit margin (%)	50.96	49.47	40.84
Net profit margin (%)	31.02	29.54	24.47
Debt ratio (%)	44.76	53.01	47.48
Current ratio	0.93	0.98	1.17
Quick ratio	0.44	0.53	0.49
Accounts receivable turnover ratio	2.73	2.60	3.74

Source Wind, compiled by ZLCA

(3) Income Structure (Table 7.107)

Table 7.107 Income structure of orient Tonghua from 2017 to 2019 (unit *yuan* in ten thousands, %)

Category	2019		2018		2017	
	Amount	Ratio	Amount	Ratio	Amount	Ratio
Engineering construction	13,162.04	68.36	6623.02	38.36	8537.99	52.36
Operation service	6090.83	31.64	3386.99	19.62	3834.38	23.51
Merchandising	–	–	6506.85	37.69	3935.09	24.13
Technical service income	–	–	749.29	4.34	–	–
Total	19,252.87	100	17,266.15	100	16,307.46	100

Source Wind, compiled by ZLCA

7.2.36 Beijing Action Information Technology Co., Ltd.

1. Basic information of the Company

Beijing Action Information Technology Co., Ltd. (hereinafter referred to as 'Action') was established in 2005, publicly listed in NEEQ (National Equities Exchange and Quotations) in 2016 (stock code: 837,950.OC). The Market value of Action is 346 million *yuan* on 31st of December 2019.

As the leading comprehensive service provider of information and communication technology in China, Action is a high-tech enterprise dedicated to providing professional telecommunication network technical services and system solutions. Its business areas include design, installation, commissioning, optimization, software development, mobile proxy and so on. Action integrates products and technical to provide technical solutions for various industries by conducting comprehensive business cooperation with leading communication mainstream manufacturers and service suppliers at home and abroad.

The company is in the 5G industry chain belonging to the information transmission, software and IT service. Along with the arrival of 5G era, the coordinated development of all industries will be driven, the current social form will be reorganized, and the intelligent and automatic upgrade of social infrastructure will be realized. The application of 5G industry will be gradually increased, with rapid development of industries such as animal networking, big data and artificial intelligence. Thanks to the development and trend of the industry, Action will be in a growth trend in the medium- and long-term time.

The customer concentration is significantly high. In 2019, the sales revenue generated by the company's top five customers accounted for 47.65, 33.09, 4.88, 2.15 and 1.61% of the operating revenue of the year, totaling 89.38%.

2. Financial Analysis

(1) Basic Financial Data (Table 7.108)

Table 7.108 Financial data for action from 2017 to 2019 (unit (*yuan* in 10 thousands)s)

	2019	2018	2017
Total assets	20,386.56	15,060.88	12,500.01
Net income attributed to shareholders	14,976.40	12,784.44	10,398.46
Revenue	34,431.96	30,519.62	25,199.92
Net profit	2191.96	1694.82	1810.08
Cash flow from operating activities	535.54	−830.63	1516.45

Source Wind, compiled by ZLCA

(2) Primary Financial Data (Table 7.109)

Table 7.109 Primary financial data for action from 2017 to 2019

	2019	2018	2017
Gross profit margin (%)	16.41	16.81	18.42
Net profit margin (%)	6.37	5.55	7.18
Debt ratio (%)	26.54	15.12	16.81
Current ratio	3.69	6.39	5.67
Quick ratio	3.69	6.39	5.49
Accounts receivable turnover ratio	3.79	6.41	6.94

Source Wind, compiled by ZLCA

(3) Income Structure (Table 7.110)

Table 7.110 Income structure for action from 2017 to 2019 (unit *yuan* in ten thousands, %)

Category	2019		2018		2017	
	Amount	Ratio	Amount	Ratio	Amount	Ratio
Communication network integration service	20,535.58	59.64	12,858.83	42.13	9124.42	36.21
Communication network optimization and consulting service	7868.34	22.85	5270.87	17.27	3603.66	14.30
Communication network security and assurance service	4602.63	13.37	10,546.89	34.56	11,164.09	44.30
Software development and system solution	1183.56	3.44	548.81	1.80	689.04	2.73
Hardware product sales	241.85	0.70	1294.21	4.24	618.71	2.46
Total	34,431.96	100	30,519.62	100	25,199.92	100

Source Wind, compiled by ZLCA

7.2.37 Beijing Kuaiyu Electronics Co., Ltd.

1. Basic information of the Company

Beijing Kuaiyu Electronics Co., Ltd. (hereinafter referred to as 'Kuaiyu') was established in 2006 publicly listed in NEEQ (National Equities Exchange and Quotations) in 2016 (stock code: 838,168.OC). The Market value of Kuaiyu is 160 million *yuan* on 31st of December 2019.

Kuaiyu is a manufacturer of computer, communication and other electronic equipment, mainly engaged in the development, production and sales of audio monitoring products and intelligent voice analysis products. Through system integrators and engineers, the company indirectly provides terminal customers with excellent performance of audio products such as pickup, intercom, recording server, broadcast conference system, etc., engaged in security audio core technology research and always keep the industry cutting-edge technology level, actively promote intelligent audio applications in various industries in recent years the ground as well.

In the field of security audio monitoring subdivision, which incorporates cross-cutting technologies such as acoustic technology, audio analysis and software development, the company keeping track of the latest technological achievements and development direction in the field of global audio, in terms of traditional security business related technologies and products to keep the r&d input at the same time, with the leading core algorithm, by intelligent audio new technology research and development and improvement of the circuit, and constantly improve the quality of products and services, strengthen study in depth, audio data application, Internet development in areas such as the foundation, on the technical innovation and product innovation continues to lead the market.

The customer concentration is fairly low. the sales revenue generated by the company's top five customers accounted for 8.49, 4.73, 4.00, 2.72 and 1.93% of the operating revenue of the year, respectively, accounting for 21.87% in total.

2. Financial Analysis

   (1) Basic Financial Data (Table 7.111)

Table 7.111 Financial data for Kuaiyu from 2017 to 2019 (unit (*yuan* in 10 thousands))

	2019	2018	2017
Total assets	9852.22	6983.20	6830.14
Net income attributed to shareholders	7210.22	5393.34	5073.51
Revenue	15,490.37	12,115.51	8471.98
Net profit	3039.40	1920.55	2779.94
Cash flow from operating activities	147.39	262.24	1451.24

Source Wind, compiled by ZLCA

(2) Primary Financial Data (Table 7.112)

Table 7.112 Primary financial data for Kuaiyu from 2017 to 2019

	2019	2018	2017
Gross profit margin (%)	55.89	59.62	78.63
Net profit margin (%)	19.62	15.85	32.81
Debt ratio (%)	26.29	22.35	25.31
Current ratio	3.52	4.24	3.77
Quick ratio	2.87	3.16	2.83
Accounts receivable turnover ratio	4.49	7.33	10.33

Source Wind, compiled by ZLCA

(3) Income Structure (Table 7.113)

Table 7.113 Income structure for Kuaiyu from 2017 to 2019 (unit *yuan* in ten thousands, %)

Category	2019		2018		2017	
	Amount	Ratio	Amount	Ratio	Amount	Ratio
Audio monitoring equipment and accessories products	15,483.54	99.96	12,115.51	100.00	8471.98	100
Other business	6.83	0.04	–	–	–	–
Total	15,490.37	100	12,115.51	100	8471.98	100

Source Wind, compiled by ZLCA

7.2.38 *Beijing JZHT Finance Securities Service Co., Ltd.*

1. Basic information of the Company

Beijing JZHT Finance Securities Service Co., LTD. (hereinafter referred to as 'JZHT') was established in 2005, publicly listed in NEEQ (National Equities Exchange and Quotations) in 2016 (stock code: 838,334.OC). The Market value of JZHT is 183 million *yuan* on 31st of December 2019.

As a comprehensive provider of capital and financial services whose main business is investor relations management services, JZHT provides customers with comprehensive online and offline services relying on the auxiliary services of market value management provided by Niuniu financial interactive platform. The business sector consists of investor relations services, securities affairs management services and consulting services for listed companies.

Investor relations management industry is not long in China's development process, and the standardization of the industry, maturity is far from the level of developed markets. With the gradual growth and maturity of the capital market and the increasing complexity and diversity of listed entities and capital markets, the status and role of the investment and customs industry will be increasingly improved, and the market scale has a large space for imagination. However, the requirements on the professional level of the industry will also be correspondingly improved. The investor relations management industry in which the company is engaged mainly serves listed (NEEQ listed) companies and companies intend to be listed (intend to be listed in NEEQ), which is closely related to the capital market and policy trend. The company's goal of being a leader in the industry with rigorous and professional services will not change, and the strategic development plan of capital service industry chain, which is committed to huge space, will not change.

The customer concentration is fairly high. the sales revenue generated by the company's top five customers accounted for 8.84, 8.19, 5.70, 4.02 and 3.55% of the operating revenue of the year respectively, with a total of 30.30%, and the company didn't disclose its top five clients in 2019.

2. Financial Analysis

(1) Basic Financial Data (Table 7.114)

Table 7.114 Financial data for JZHT from 2017 to 2019 (unit (*yuan* in 10 thousands))

	2019	2018	2017
Total assets	13,110.26	8999.18	8411.92
Net income attributed to shareholders	8193.34	5776.48	5110.85
Revenue	7585.22	5893.48	10,758.74
Net profit	3518.46	2266.44	2507.24
Cash flow from operating activities	1405.75	1806.62	3416.10

Source Wind, compiled by ZLCA

(2) Primary Financial Data (Table 7.115)

Table 7.115 Primary financial data for JZHT from 2017 to 2019

	2019	2018	2017
Gross profit margin (%)	53.50	53.10	49.76
Net profit margin (%)	46.39	38.46	23.30
Debt ratio (%)	37.50	35.81	39.24
Current ratio	2.30	2.75	2.11
Quick ratio	2.30	2.75	2.11
Accounts receivable turnover ratio	6.77	6.33	10.06

Source Wind, compiled by ZLCA

(3) Income Structure (Table 7.116)

Table 7.116 Income structure for JZHT from 2017 to 2019 (unit *yuan* in ten thousands, %)

Category	2019		2018		2017	
	Amount	Ratio	Amount	Ratio	Amount	Ratio
Service	6305.74	83.13	4880.59	82.81	10,408.97	96.75
Perennial service	1279.48	16.87	1012.89	17.19	349.76	3.25
Total	7585.22	100	5893.48	100	10,758.74	100

Source Wind, compiled by ZLCA

7.2.39 Beijing Bang Yuan Environment Science and Technology Limited Co., Ltd.

1. Basic information of the Company

Beijing Bang Yuan environment science and technology limited co., Ltd. (hereinafter referred to as 'Bang Yuan') was established in 2008, publicly listed in NEEQ (National Equities Exchange and Quotations) in 2016 (stock code: 838,740.OC). The Market value of Bang Yuan is 5 million *yuan* on 31st of December 2019.

The company is a national high-tech enterprise specializing in the ecological restoration of black and smelly, eutrophic organisms of surface water (river course, lake). The main business of the company is the construction and related design services of water environment bio-ecological restoration project, including urban river ecological management, park water environmental ecological management, lake ecological management, landscape water ecological management, etc.

Water treatment industry are incompletely known, so there may be risks of unsuccessful expansion of new customers, such as backward concept of water treatment in new regions, differences in climate and geographical environment, etc. However, with the encouragement of national policies, the new market development of the company also presents a wider range of opportunities and demands, and the company will promote the development of new regions and new markets under the business of the inherent region.

The customer concentration is fairly high. In 2019, the sales revenue generated by the company's top five customers accounted for 18.88, 15.39, 7.84, 7.69, 6.89% of the operating revenue of the year, respectively, totaling 56.69%.

2. Financial Analysis

(1) Basic Financial Data (Table 7.117)

Table 7.117 Financial data for Bang Yuan from 2017 to 2019 (unit (*yuan* in 10 thousands))

	2019	2018	2017
Total assets	7729.64	6156.35	3648.33
Net income attributed to shareholders	6147.57	3684.99	2346.24
Revenue	9053.20	7679.66	4907.59
Net profit	2762.59	2338.75	1448.83
Cash flow from operating activities	−1240.80	3635.22	1214.43

Source Wind, compiled by ZLCA

(2) Primary Financial Data (Table 7.118)

Table 7.118 Primary financial data for Kuaiyu from 2017 to 2019

	2019	2018	2017
Gross profit margin (%)	49.81	46.95	53.95
Net profit margin (%)	30.52	30.45	29.52
Debt ratio (%)	20.47	40.14	35.69
Current ratio	4.82	2.43	2.67
Quick ratio	2.14	2.01	2.56
Accounts receivable turnover ratio	19.89	8.37	6.23

Source Wind, compiled by ZLCA

(3) Income Structure (Table 7.119)

Table 7.119 Income structure for Kuaiyu from 2017 to 2019 (unit *yuan* in ten thousands, %)

Category	2019		2018		2017	
	Amount	Ratio	Amount	Ratio	Amount	Ratio
Bio-ecological water treatment project	6695.69	73.96	6018.43	78.37	3738.62	76.18
Design service	1697.87	18.75	1095.45	14.26	767.70	15.64
Other business	659.64	7.29	565.77	7.37	401.27	8.18
Total	15,490.37	100	12,115.51	100	8471.98	100

Source Wind, compiled by ZLCA

7.2.40 *Beijing SG-Culture Media Corp., Ltd.*

1. Basic information of the Company

Beijing SG-Culture Media Corp., Ltd (hereinafter referred to as 'SG-Culture') was established in 2013, publicly listed in NEEQ (National Equities Exchange and Quotations) in 2016 (stock code: 839,463.OC). The Market value of SG-Culture is 704 million *yuan* on 31st of December 2019.

The company focuses on the creation of TV series. Main business is the investment, production and distribution of TV dramas and TV programs. Among them, TV drama copyright income is the main source of the company's income. The company's customer types mainly include TV stations, agency distribution companies, network platforms, film and television companies.

The production regulation of the TV program industry in which the company is located has gradually changed from government-oriented to market-oriented and has begun to realize the survival of the fittest through market competition. At present, the television program industry policy entry threshold is low, the production organization number is numerous, the market concentration is not high, the competition is quite sufficient.

The customer concentration is fairly high, the ratios as the each of top 5 customers sales by the total revenue are 17.55, 15.21, 12.64, 11.08, 10.70%, totaling 67.18%. The company doesn't reveal the specific names of the top 5 customers in 2019.

2. Financial Analysis

(1) Basic Financial Data (Table 7.120)

Table 7.120 Financial data for SG-Culture from 2017 to 2019 (unit (*yuan* in 10 thousands))

	2019	2018	2017
Total assets	52,643.84	60,327.51	39,354.58
Net income attributed to shareholders	27,526.12	24,446.13	20,567.76
Revenue	16,770.98	36,760.79	24,175.02
Net profit	3079.99	3878.37	4590.35
Cash Flow from Operating Activities	−1032.69	5784.75	−9323.58

Source Wind, compiled by ZLCA

(2) Primary Financial Data (Table 7.121)

Table 7.121 Primary financial data for SG-Culture from 2017 to 2019

	2019	2018	2017
Gross profit margin (%)	55.15	54.88	44.26
Net profit margin (%)	18.37	10.55	18.99
Debt ratio (%)	47.71	59.48	47.74
Current ratio	2.08	1.69	2.16
Quick ratio	1.08	1.15	1.39
Accounts receivable turnover ratio	0.69	1.41	1.43

Source Wind, compiled by ZLCA

(3) Income Structure (Table 7.122)

Table 7.122 Income structure for SG-culture from 2017 to 2019 (unit *yuan* in ten thousands, %)

Category	2019		2018		2017	
	Amount	Ratio	Amount	比例	Amount	Ratio
Teleplay business	7851.16	46.81	33,692.65	91.65	17,058.19	70.56
Daigou agent program business	7293.63	43.49	1232.95	3.35	3128.94	12.94
Co-production proceeds	1539.87	9.18	–	–	753.21	3.12
Variety show and planning services	86.32	0.52	1728.58	4.70	1397.26	5.78
Produced a trustee business	–	–	–	–	1698.11	7.02
Advertising business	–	–	106.60	0.30	139.31	0.58
Total	16,770.98	100.00	36,760.78	100.00	24,175.02	100.00

Source Wind, compiled by ZLCA

7.2.41 Yonyou Fintech Technology Co., Ltd.

1. Basic information of the Company

Yonyou Fintech Technology Co., Ltd. (hereinafter referred to as 'Yonyou Fintech') was established in 2004, publicly listed in NEEQ (National Equities Exchange and Quotations) in 2016 (stock code: 839,483.OC). The Market value of Yonyou Fintech is 1.021 billion *yuan* on 31st of December 2019.

The company provides consulting, software, industry solutions and professional services to banking, securities, trust and other financial enterprises. The company's main business covers financial enterprise operation and management, business transactions, marketing services, business analysis and other fields, while providing training, operation and maintenance, data, outsourcing services.

Company's financial industry will be along with the deepening reform of financial supply side, from extensive supply of indifference to "science and technology leading the financial" as the core of fine regulation of pay-as-you-go new stage, will build by means of "financial + scene" financial ecosystem, to achieve product line, business platform, digital and ecological service, meet the demand of customer integration.

The customer concentration is fair low, the ratios as the each of top 5 customers sales by the total revenue are 2.52, 2.47, 2.33, 2.31, 1.94%, totaling 11.57%.

2. Financial Analysis

(1) Basic Financial Data (Table 7.123)

Table 7.123 Financial data for Yonyou fintech from 2017 to 2019 (unit (*yuan* in 10 thousands))

	2019	2018	2017
Total Assets	37,930.05	30,251.03	25,519.98
Net income attributed to shareholders	27,212.84	22,022.31	16,699.21
Revenue	38,069.26	34,383.43	33,310.01
Net Profit	6157.71	4846.52	2083.87
Cash Flow from Operating Activities	7620.04	4222.09	−4768.37

Source Wind, compiled by ZLCA

(2) Primary Financial Data (Table 7.124)

Table 7.124 Primary financial data for Yonyou fintech from 2017 to 2019

	2019	2018	2017
Gross profit margin (%)	62.78	60.58	56.12
Net profit margin (%)	16.18	14.10	6.26
Debt ratio (%)	28.26	27.20	34.56
Current ratio	2.72	2.59	2.34
Quick ratio	2.72	2.59	2.34
Accounts receivable turnover ratio	3.18	3.00	3.32

Source Wind, compiled by ZLCA

(3) Income Structure (Table 7.125)

Table 7.125 Income structure for Yonyou fintech from 2017 to 2019 (unit *yuan* in ten thousands, %)

Category	2019		2018		2017	
	Amount	Ratio	Amount	Ratio	Amount	Ratio
Bank	7851.16	46.81	33,692.65	91.65	17,058.19	70.56
Securities, funds, futures	7293.63	43.49	1232.95	3.35	3128.94	12.94
Other financial industries	1539.87	9.18	–	–	753.21	3.12
Insurance	86.32	0.52	1728.58	4.70	1397.26	5.78
Lease	–	–	–	–	1698.11	7.02
Trust	–	–	106.60	0.30	139.31	0.58
Total	16,770.98	100.00	36,760.78	100.00	24,175.02	100.00

Source Wind, compiled by ZLCA

7.2.42 Trustwin Communication Brand Consulting (Beijing) Co., Ltd.

1. Basic information of the Company

Trustwin Communication Brand Consulting (Beijing) Co.,Ltd (hereinafter referred to as 'Trustwin Brand') was established in 2010, publicly listed in NEEQ (National Equities Exchange and Quotations) in 2016 (stock code: 839,622.OC). The Market value of Trustwin Brand is 360 million *yuan* on 31st of December 2019.

The company is a brand marketing service provider that helps customers enhance brand influence and brand value. The main business of the company is to provide integrated marketing communication services, which can be divided into brand promotion services for the overall brand image of customers and marketing services for specific products of customers.

Companies in the industry to participate in the competition of various subjects, the overall concentration of the market is not high, the industry competition is fierce. With the improvement of the standardization of the public relations industry, the competition among domestic public relations companies will become increasingly fierce, and companies with small resources and weak strength will be gradually eliminated by the market. Trustwin Brand is facing competitive pressures from local and other parts of the country and is at risk of increased competition in the market for existing products and services.

The customer concentration is significantly high, the ratios as the each of top 5 customers sales by the total revenue are 24.65%, 21.07%, 14.65%, 8.55%, 3.96% respectively, totaling 72.88%. The company doesn't reveal the specific names of the top 5 customers in 2019.

2. Financial Analysis

(1) Basic Financial Data (Table 7.126)

Table 7.126 Financial data for Trustwin brand from 2017 to 2019 (unit (*yuan* in 10 thousands))

	2019	2018	2017
Total assets	24,662.70	17,227.26	9291.83
Net income attributed to shareholders	13,378.18	11,061.34	4273.85
Revenue	33,403.81	26,795.83	16,442.64
Net profit	2750.62	2911.52	2161.99
Cash flow from operating activities	1083.33	−3404.53	464.84

Source Wind, compiled by ZLCA

(2) Primary Financial Data (Table 7.127)

Table 7.127 Primary financial data for Trustwin brand from 2017 to 2019

	2019	2018	2017
Gross profit margin (%)	21.46	24.53	32.71
Net profit margin (%)	8.23	10.87	13.15
Debt ratio (%)	44.71	33.55	50.97
Current ratio	2.21	2.96	1.96
Quick ratio	1.84	2.21	1.15
Accounts receivable turnover ratio	2.63	3.73	3.60

Source Wind, compiled by ZLCA

(3) Income Structure (Table 7.128)

Table 7.128 Income structure for Trustwin brand from 2017 to 2019 (unit *yuan* in ten thousands, %)

Category	2019		2018		2017	
	Amount	Ratio	Amount	Ratio	Amount	Ratio
Marketing service	33,403.81	100.00	26,795.83	100.00	16,442.64	100.00
Total	33,403.81	100.00	26,795.83	100.00	16,442.64	100.00

Source Wind, compiled by ZLCA

7.2.43 Know-How Education Ltd.

1. Basic information of the Company

Know-how Education Ltd. (hereinafter referred to as 'Know-how') was established in 2003, publicly listed in NEEQ (National Equities Exchange and Quotations) in 2016 (stock code: 839,994.OC). The Market value of Know-how is 515 million *yuan* on 31st of December 2019.

The company is an international, professional, standardized vocational education curriculum system providers and service providers. The main business of the company is to provide the curriculum system of automobile vocational education and the systematic solution of the training of high-skilled automobile talents for the national middle and higher vocational colleges, and to provide students with automobile vocational education services in line with international standards and higher quality employment services.

With the rapid development of the education industry where the company is located, the demand for high-skilled talents is increasing day by day, and the auxiliary education industry aimed at middle and higher vocational colleges will also get rapid development. As more and more enterprises enter the industry, the competition will gradually intensify. The homogenized competition of educational value-added service business is inevitable, and the company will face the risk of intensified market competition in the future.

The customer concentration is fairly low, the ratios as the each of top 5 customers sales by the total revenue are 3.55%,1.48%,1.16%,1.04%,0.98%, totaling 8.21%. The company doesn't reveal the specific names of the top 5 customers in 2019.

2. Financial Analysis

   (1) Basic Financial Data (Table 7.129)

Table 7.129 Financial data for Know-how from 2017 to 2019 (unit (*yuan* in 10 thousands))

	2019	2018	2017
Total assets	14,280.58	15,622.69	13,998.35
Net income attributed to shareholders	9606.76	9845.07	8697.42
Revenue	9079.49	8489.75	7835.02
Net profit	1760.29	1551.62	1566.45
Cash flow from operating activities	499.67	2092.34	2324.13

Source Wind, compiled by ZLCA

(2) Primary Financial Data (Table 7.130)

Table 7.130 Primary financial data for Know-how from 2017 to 2019

	2019	2018	2017
Gross profit margin (%)	56.49	50.75	54.30
Net profit margin (%)	19.39	18.28	19.99
Debt ratio (%)	32.73	36.98	37.87
Current ratio	2.10	2.11	2.07
Quick ratio	2.07	2.10	2.06
Accounts receivable turnover ratio	13.66	22.04	26.79

Source Wind, compiled by ZLCA

(3) Income Structure (Table 7.131)

Table 7.131 Income structure for know-how from 2017 to 2019 (unit *yuan* in ten thousands, %)

Category	2019		2018		2017	
	Amount	Ratio	Amount	Ratio	Amount	Ratio
Project service	8195.10	90.26	8123.75	95.69	7812.24	99.71
Product sales	857.21	9.44	336.76	3.97		–
Other business	27.18	0.30	29.25	0.34	22.78	0.29
Total	9079.49	100.00	8489.76	100.00	7835.02	100.00

Source Wind, compiled by ZLCA

7.2.44 Beijing 797 Audio Co., Ltd.

1. Basic information of the Company

Beijing 797 Audio Co., Ltd. (hereinafter referred to as '797') was established in 1994, publicly listed in NEEQ (National Equities Exchange and Quotations) in 2016 (stock code: 870,381.OC). The Market value of 797 is 821 million *yuan* on 31st of December 2019.

'797' is a large electroacoustic enterprise dedicated to providing intelligent acoustic integrated solutions. The company's main business includes the development and production of professional audio products such as microphones, capacitive microphones and loudspeakers, which are widely used in culture, sports, conference systems, public broadcasting and other fields, and are exported to the United States, Germany, Australia, Japan, Brazil and other countries.

In the environment where manufacturing industry competition continues to intensify, intelligent acoustics is in a state of rapid growth, and its product intelligent speakers will be more and more supported by the market due to their highly integrated functions and powerful voice interaction capabilities. The company also attaches more and more importance to intelligent acoustics, and in accordance with the established business plan, the steady development of the business.

The customer concentration is fairly high, the ratios as the each of top 5 customers sales by the total revenue are 18.64, 17.03, 10.34, 9.04, 2.66%, totaling 57.71%.

2. Financial Analysis

(1) Basic Financial Data (Table 7.132)

Table 7.132 Financial data for 797 from 2017 to 2019 (unit (*yuan* in 10 thousands))

	2019	2018	2017
Total assets	32,013.46	29,320.43	23,674.71
Net income attributed to shareholders	23,563.43	19,435.11	16,789.73
Revenue	30,622.26	27,937.89	26,555.84
Net profit	6275.05	4073.92	3330.62
Cash flow from operating activities	2105.62	1802.12	648.77

Source Wind, compiled by ZLCA

(2) Primary Financial Data (Table 7.133)

Table 7.133 Primary financial data for 797 from 2017 to 2019

	2019	2018	2017
Gross profit margin (%)	34.41	25.88	29.70
Net profit margin (%)	20.49	14.58	12.54
Debt ratio (%)	24.89	32.23	27.36
Current ratio	3.99	2.93	3.42
Quick ratio	3.10	2.26	2.73
Accounts receivable turnover ratio	3.36	4.32	7.99

Source Wind, compiled by ZLCA

(3) Income Structure (Table 7.134)

Table 7.134 Income structure for 797 from 2017 to 2019 (unit *yuan* in ten thousands, %)

Category	2019		2018		2017	
	Amount	Ratio	Amount	Ratio	Amount	Ratio
ODM	16,391.03	53.53	20,471.84	73.28	17,529.95	66.01
Intelligent acoustic engineering and technical services	11,283.17	36.85	4914.89	17.59	5111.39	19.25
Professional audio products	1540.28	5.03	1158.77	4.15	2195.39	8.27
OEM	1407.78	4.60	1391.61	4.98	1689.12	6.36
Other business			0.78	0.00	29.99	0.11
Total	30,622.26	100.00	27,937.89	100.00	26,555.84	100.00

Source Wind, compiled by ZLCA

7.2.45 *Beijing ZBX Information Technology Co., Ltd.*

1. Basic information of the Company

Beijing ZBX Information Technology Co., Ltd. (hereinafter referred to as "ZBX") was established in 2002, publicly listed in NEEQ (National Equities Exchange and Quotations) in 2017 (stock code: 870,992.OC). The Market value of ZBX is 183.95 million *yuan* on 31st of December 2019.

The company is China's information consulting services and information systems engineering supervision service providers. Company's main business information system engineering supervision as the core, covering the informationization construction planning consultation, project evaluation consultation, preliminary design, project management consulting, project supervision, performance evaluation, evaluation after consulting IT business chain, to provide users with comprehensive supervision in the field of information engineering services and related consulting services. The company's main services are classified into two categories: 1. Information system engineering supervision services; 2. Information consulting services.

The information system engineering supervision industry where the company is located is an emerging industry. With the continuous deepening and development of national information construction, technologies such as cloud computing and Internet of Things have been widely applied, and the national software and information technology service industry is still showing a rapid growth trend. With the need of national security, IT provides a huge developable market for localized information consulting services, which lays a foundation for IT enterprises to achieve rapid development in the field of information consulting services.

The customer concentration is fairly low, and the sales revenue of the company's top five customers accounted for 2.7, 2.65, 2.24, 1.74 and 1.72% of the operating revenue of the year respectively, with a total of 11.05%. The company didn't disclose its top five customers in 2019.

2. Financial Analysis

   (1) Basic Financial Data (Table 7.135)

Table 7.135 Financial data for ZBX from 2017 to 2019 (unit (*yuan* in 10 thousands))

	2019	2018	2017
Total assets	14,579.80	12,875.23	8480.05
Net income attributed to shareholders	12,585.33	10,449.61	6116.38
Revenue	13,135.18	11,165.61	7709.22
Net profit	2274.90	3111.03	1568.76
Cash flow from operating activities	1104.67	1842.14	−112.12

Source Wind, compiled by ZLCA

(2) Primary Financial Data (Table 7.136)

Table 7.136 Primary financial data for ZBX from 2017 to 2019

	2019	2018	2017
Gross profit margin (%)	52.19	57.86	52.77
Net profit margin (%)	17.32	27.86	20.35
Debt ratio (%)	13.68	18.84	27.87
Current ratio	7.12	5.20	3.50
Quick ratio	7.12	5.20	3.50
Accounts receivable turnover ratio	2.20	2.36	2.62

Source Wind, compiled by ZLCA

(3) Income Structure (Table 7.137)

Table 7.137 Income structure for ZBX from 2017 to 2019 (unit *yuan* in ten thousands, %)

Category	2019		2018		2017	
	Amount	Ratio	Amount	Ratio	Amount	Ratio
Supervising business	11,951.95	90.99	10,584.58	94.80	7290.87	94.57
Consulting business	1183.23	9.01	581.04	5.20	409.83	5.32
Total	13,135.18	100.00	11,165.61	100.00	7700.70	99.89

Source Wind, compiled by ZLCA

7.2.46 Sugon Data Energy (Beijing) Co., Ltd.

1. Basic information of the Company

Sugon Data Energy (Beijing)Co., Ltd. (hereinafter referred to as "Sugon Energy") was established in 2002, publicly listed in NEEQ (National Equities Exchange and Quotations) in 2016 (stock code: 872,808.OC). The Market value of Sugon Energy is 330 million *yuan* on 31st of December 2019.

Sugon Energy is a high efficient refrigeration technology as the core of data center overall solutions provider, is a highly efficient cooling technology as the core of data center overall solutions provider, is mainly engaged in liquid cooling products, computer room air conditioning products, room temperature and humidity control products, equipment room micro module product research, development, sales, and around the above products provide software development, system integration and technical services, is a leader in the field of data center of China refrigeration.

The company is in the data center energy saving industry with the explosive growth of the Internet, the Internet of Things, cloud computing and other data business and the rapid development of IT technology and rapid growth. In the future, China's data center market will still have a strong driving force, while the infrastructure requirements closely related to the data center will gradually increase. Under the situation of server products becoming more and more standardized, the efficient and energy-saving supporting infrastructure will gradually become a key factor to attract customers.

The customer concentration is significantly high, the sales revenue generated by the company's top five customers accounted for 77.68, 3.65, 3.16, 2.63 and 1.31% of the operating revenue of the year, respectively, with a total of 88.43%.

2. Financial Analysis

(1) Basic Financial Data (Table 7.138)

Table 7.138 Financial data for sugon energy from 2017 to 2019 (unit (*yuan* in 10 thousands))

	2019	2018	2017
Total assets	17,413.16	16,655.89	12,087.67
Net income attributed to shareholders	7025.77	4509.11	3682.01
Revenue	29,170.87	15,920.18	11,343.06
Net profit	2819.81	1003.79	560.41
Cash flow from operating activities	547.07	−602.09	2735.71

Source Wind, compiled by ZLCA

(2) Primary Financial Data (Table 7.139)

Table 7.139 Primary financial data for sugon energy from 2017 to 2019

	2019	2018	2017
Gross profit margin (%)	26.29	24.68	19.85
Net profit margin (%)	9.67	6.31	4.94
Debt ratio (%)	59.65	72.93	69.54
Current ratio	1.51	1.21	1.19
Quick ratio	0.49	0.74	0.51
Accounts receivable turnover ratio	11.53	8.27	23.30

Source Wind, compiled by ZLCA

(3) Income Structure (Table 7.140)

Table 7.140 Income structure for sugon energy from 2017 to 2019 (unit *yuan* in ten thousands, %)

Category	2019		2018		2017	
	Amount	Ratio	Amount	Ratio	Amount	Ratio
Liquid-cooled product	23,813.03	81.63	11,784.27	74.02	6134.20	54.08
Modular data center product	5312.94	18.21	4037.86	25.36	3950.24	34.83
Technical and operational service	44.90	0.15	98.05	0.62	1258.62	11.09
Total	29,170.87	100.00	15,920.18	100.00	11,343.06	100.00

Source Wind, compiled by ZLCA

7.2.47 Beijing Jiuzhoudadi Bio-Tech Group Co., Ltd.

1. Basic information of the Company

Beijing Jiuzhoudadi Bio-Tech Group Co., Ltd. (hereinafter referred to as "dadi Ltd.") was established in 2000, publicly listed in NEEQ (National Equities Exchange and Quotations) in 2016 (stock code: 430,034.OC). The Market value of dadi Ltd is 124 *yuan* on 31st of December 2019.

The company is the domestic first-class feed production enterprises. The products are premix, protein concentrate, mixed feed and animal medicine of various kinds of livestock, poultry, aquatic products, ruminants and economic animals, which are suitable for customers' needs and with the best cost performance. There are as many as hundreds of varieties, and the technical indexes of the products are in the leading position in China and reach the international advanced level. The company enjoys the preferential tax policy of 15% enterprise income tax rate. Its products have high scientific and technological content, high added value, high market share and strong competitiveness.

The feed industry in which the company is located is the most industrialized industry in China's agriculture. It is in the middle link of the industrial chain of "planting industry—feed industry—breeding industry", which is not only affected by the double pressure at the front and the back end, but also affects their development. Feed industry benefits planting industry and breeding industry, which is an important pillar industry of China's rural economy. The vigorous development of feed industry not only provides the material basis for the breeding industry, but also plays a vital role in ensuring food safety and hygiene. Feed industry is driven by domestic demand to develop an industry, in the long run, feed industry has a relatively broad development prospects, a large room for market expansion.

The concentration of the company's customers is fairly low. In 2019, the sales revenue generated by the company's top five customers accounted for 6.72, 3.47,

2.47, 1.63 and 1.52% of the operating revenue of the year, respectively, with a total of 15.81%.

2. Financial Analysis

(1) Basic Financial Data (Table 7.141)

Table 7.141 Financial data for dadi Ltd. from 2017 to 2019 (unit (*yuan* in 10 thousands))

	2019	2018	2017
Total assets	101,754.51	95,834.02	98,200.44
Net income attributed to shareholders	59,364.56	53,923.28	27,431.97
Revenue	165,309.93	150,569.74	133,314.19
Net profit	8088.26	5838.80	6929.00
Cash flow from operating activities	10,263.10	6754.25	1712.82

Source Wind, compiled by ZLCA

(2) Primary Financial Data (Table 7.142)

Table 7.142 Primary financial data for dadi Ltd. from 2017 to 2019

	2019	2018	2017
Gross profit margin (%)	16.55	17.54	19.93
Net profit margin (%)	4.89	3.88	5.20
Debt ratio (%)	34.25	34.29	37.69
Current ratio	1.83	1.62	1.37
Quick ratio	1.07	0.89	0.71
Accounts receivable turnover ratio	17.67	18.12	18.86

SourceWind, compiled by ZLCA

(3) Income Structure (Table 7.143)

Table 7.143 Income structure for dadi Ltd. from 2017 to 2019 (unit *yuan* in ten thousands, %)

Category	2019		2018		2017	
	Amount	Ratio	Amount	Ratio	Amount	Ratio
Feed	137,917.04	83.43	123,001.76	81.69	103,795.72	77.86
Milk and others	21,350.12	12.92	22,416.43	14.89	24,117.41	18.09
Other business	2877.00	1.74	1943.31	1.29	2167.88	1.63
Biological assets and frozen essence	900.92	0.54	1880.06	1.25	1996.57	1.50
Technical service fee	45.31	0.03	986.05	0.65	0.00	0.00
Total	165,309.93	100.00	150,569.74	100.00	132,077.58	99.07

Source Wind, compiled by ZLCA

7.2.48 *Meta Space (Beijing) Air Dome Corp.*

1. Basic information of the Company

Meta Space (Beijing) Air Dome Corp. (hereinafter referred to as “Meta Space Air Dome”) was established in 2006, publicly listed in NEEQ (National Equities Exchange and Quotations) in 2016 (stock code: 831,527.OC). The Market value of Meta Space Air Dome is 122 million *yuan* on 31st of December 2019.

Meta Space Air Dome is committed to promoting green and environmentally friendly gas film architecture in the Chinese market, spreading advanced architectural concepts and displaying new life experience. Based on the domestic market, the company has researched and developed a number of gas film building technology patents, with first-class design ability, all-round service ability and rich construction experience. According to the needs of different industries, different types of customers to provide professional film solutions. At present, meta space air dome has been put into use in many cities and many different fields in China.

The end of the construction industry in which the company is in a period of rapid development, and the improvement of the people’s material and cultural level leads to the demand for various sports and entertainment venues; At the same time, the rapid development of industrialization has also brought more and more prominent problems of air pollution, the state began to pay attention to pollution control from the source, closed pollutant treatment buildings are gradually being widely used in the field of industrial environmental protection. At present, the industry concentration is high and the industry competition is low.

The customer concentration is fairly high, the ratio as the each of top 5 customers sales by the total revenue are 7.88%, 7.68%, 6.04%, 5.86%, 5.00%, totaling 32.46%. The company doesn’t reveal the specific names of the top 5 customers in 2019.

2. Financial Analysis

   (1) Basic Financial Data (Table 7.144)

Table 7.144 Financial data for meta space air dome from 2017 to 2019 (unit (*yuan* in 10 thousands))

	2019	2018	2017
Total assets	41,786.25	28,412.79	24,263.38
Net income attributed to shareholders	22,322.06	19,767.03	20,853.71
Revenue	18,789.20	8792.28	12,769.81
Net profit	2573.93	1082.95	2019.69
Cash flow from operating activities	2396.96	−807.88	−2290.66

Source Wind, compiled by ZLCA

(2) Primary Financial Data (Table 7.145)

Table 7.145 Primary financial data for meta space air dome from 2017 to 2019

	2019	2018	2017
Gross profit margin (%)	55.15	54.88	44.26
Net profit margin (%)	18.37	10.55	18.99
Debt ratio (%)	47.71	59.48	47.74
Current ratio	2.08	1.69	2.16
Quick ratio	1.08	1.15	1.39
Accounts receivable turnover ratio	0.69	1.41	1.43

Source Wind, compiled by ZLCA

(3) Income Structure (Table 7.146)

Table 7.146 Income structure for meta space air dome from 2017 to 2019 (unit *yuan* in ten thousands, %)

Category	2019		2018		2017	
	Amount	Ratio	Amount	Ratio	Amount	Ratio
Air dome construction business	18,113.52	96.40	8232.10	93.63	12,690.08	99.38
Sports operation	590.31	3.14	539.20	6.13	79.73	0.62
Other business	85.37	0.45	20.98	0.24	0.00	0.00
Total	18,789.20	100.00	8792.28	100.00	12,769.81	100.00

Source Wind, compiled by ZLCA

7.2.49 Beijing Kai-De Quartz Co., Ltd.

1. Basic information of the Company

Beijing Kai-De Quartz Co., Ltd (hereinafter referred to as "Kai-De Quartz") was established in 1997, publicly listed in NEEQ (National Equities Exchange and Quotations) in 2016 (stock code: 835,179.OC). The Market value of Kai-De Quartz is 351 million *yuan* on 31st of December 2019.

The company is the leading quartz glass products processing enterprises in China, the main business for quartz instrument, quartz pipe, quartz boat and other quartz glass products research and development, production and sales. The company has a long-term accumulation and precipitation in quartz glass processing technology. In the field of quartz glass products for semiconductor integrated circuit chips, Kai-De Quartz is one of the few domestic enterprises that can process matching quartz glass products for 8-inch and 12-inch chip production line, main products include quartz instruments, quartz pipes, quartz boats and other quartz glass products.

The quartz glass products industry in which the company is belongs to the category of new materials industry, which is one of the seven strategic emerging industries in China. The new materials technology in the strategic emerging industries has become an important basis for China's economic development, the key development field of electronic information technology, modern industry and aerospace engineering as well. The state attaches great importance to the development of the new material industry and has issued a series of policies to promote the development of the new material industry. The quartz glass products produced by the company are used as raw materials in the downstream industry, which are mainly used in semiconductor integrated circuit, LED new electric light source, aerospace and aviation, chemical industry and other subsectors. Therefore, the market size of this industry is directly affected by the demand of the downstream industry.

The customer concentration of Kai-De Quartz is fairly high, the ratio as the each of top 5 customers sales by the total revenue are 30.65, 17.17, 6.77, 3.24, 2.78%, totaling 32.46%.

2. Financial Analysis

(1) Basic Financial Data (Table 7.147)

Table 7.147 Financial data for KaiDe quartz from 2017 to 2019 (unit (*yuan* in 10 thousands))

	2019	2018	2017
Total assets	21,263.50	17,333.11	14,216.65
Net income attributed to shareholders	18,446.77	15,200.74	12,038.53
Revenue	15,450.48	11,119.97	9136.04
Net profit	3246.04	2140.97	1227.98
Cash flow from operating activities	408.42	980.75	133.78

Source Wind, compiled by ZLCA

(2) Primary Financial Data (Table 7.148)

Table 7.148 Primary financial data for KaiDe quartz from 2017 to 2019

	2019	2018	2017
Gross profit margin (%)	44.26	44.97	34.90
Net profit margin (%)	21.01	19.25	13.44
Debt ratio (%)	13.25	12.30	15.32
Current ratio	5.81	5.81	4.53
Quick ratio	4.06	3.46	3.52
Accounts receivable turnover ratio	5.00	3.64	3.50

Source Wind, compiled by ZLCA

(3) Income Structure (Table 7.149)

Table 7.149 Income structure for KaiDe quartz from 2017 to 2019 (unit *yuan* in ten thousands, %)

Category	2019		2018		2017	
	Amount	Ratio	Amount	Ratio	Amount	Ratio
Quartz glass products for semiconductor integrated circuit chips	8012.69	51.86	8539.66	76.80	5823.95	63.75
Quartz products for photovoltaic solar industry	7393.76	47.85	2537.68	22.82	–	–
Quartz glass products for LED gallium arsenide industry	–	–	–	–	2445.64	26.77
Quartz glass products for aerospace and national defense	–	–	–	–	565.34	6.19
Quartz glass products for other industries	44.03	0.28	42.63	0.38	226.11	2.47
Quartz glass products used in chemical industry	–	–	–	–	75.00	0.82
Total	15,450.48	100.00	11,119.97	100.00	9136.04	100.00

Source Wind, compiled by ZLCA

7.2.50 *Brand Union (Beijing) Consulting Co. Ltd.*

1. Basic information of the Company

Brand Union (Beijing) Consulting Co. Ltd. (hereinafter referred to as "Brand Union") was established in 2002, publicly listed in NEEQ (National Equities Exchange and Quotations) in 2016 (stock code: 837,940.OC). The Market value of Brand Union is 127 million *yuan* on 31st of December 2019.

Brand Union is an enterprise who help customers to build, maintain systems and enhance brand value. specialized in providing brand communication, promotion, consulting and other public relations services. One of the company's main businesses is brand communication, promotion, consulting related activities management services.

The public relations industry develops rapidly, and the operating income grows year by year. With the continuous growth of China's economy and the rapid improvement of its international status, the demand for public relations services has been constantly stimulated, and Chinese enterprises have been deeply aware of the importance of professional public relations services. The public relations industry will continue to develop at a high speed. Brand Union will continue to strengthen the integration and transformation efforts: from large to strong, from super-large cities to other central cities, from a single service to integrated service development.

The customer concentration of company is fairly high. The sales revenue generated by the company's top five customers accounted for 26.18%, 23.75%, 2.74%, 2.23% and 2.1%, respectively, accounting for 57% of the operating revenue of the year.

2. Financial Analysis

(1) Basic Financial Data (Table 7.150)

Table 7.150 Financial data for brand union from 2017 to 2019 (unit (*yuan* in 10 thousands))

	2019	2018	2017
Total assets	14,821.23	6588.97	7087.66
Net income attributed to shareholders	6366.93	3973.25	3073.63
Revenue	16,401.96	11,410.00	7064.23
Net profit	2512.85	1000.85	1357.93
Cash flow from operating activities	2590.36	−753.86	3447.70

Source Wind, compiled by ZLCAnn

(2) Primary Financial Data (Table 7.151)

Table 7.151 Primary financial data for brand union from 2017 to 2019

	2019	2018	2017
Gross profit margin (%)	25.65	20.60	39.88
Net profit margin (%)	15.32	8.77	19.22
Debt ratio (%)	56.99	39.57	56.62
Current ratio	1.75	2.43	1.68
Quick ratio	1.57	2.17	1.66
Accounts receivable turnover ratio	2.86	15.95	10.94

Source Wind, compiled by ZLCA

(3) Income Structure (Table 7.152)

Table 7.152 Income structure for brand union from 2017 to 2019 (unit *yuan* in ten thousands, %)

Category	2019		2018		2017	
	Amount	Ratio	Amount	Ratio	Amount	Ratio
Brand communication	9689.51	59.08	9450.05	82.82	4519.31	63.97
Brand activity	6673.24	40.69	1910.97	16.75	2471.07	34.98
BMBA	39.21	0.24	48.99	0.43	73.84	1.05
Total	16,401.96	100.00	11,410.00	100.00	7064.22	100.00

Source Wind, compiled by ZLCAnn

The manufacturer's authorised representative in the EU is Springer Nature Customer Service Centre GmbH, Europaplatz 3, 69115 Heidelberg, Germany. If you have any concerns regarding our products, please contact ProductSafety@springernature.com

Printed and bound by CPI Group (UK) Ltd, Croydon, CR0 4YY
30/09/2026
02225293-0002